Repetitive Motion Planning and Control of Redundant Robot Manipulators

Yunong Zhang • Zhijun Zhang

Repetitive Motion Planning and Control of Redundant Robot Manipulators

Springer

Yunong Zhang
Zhijun Zhang
Sun Yat-sen University
Guangzhou
People's Republic of China

ISBN 978-3-642-37517-0 ISBN 978-3-642-37518-7 (eBook)
DOI 10.1007/978-3-642-37518-7
Springer Heidelberg New York Dordrecht London

Library of Congress Control Number: 2013938762

Printed on acid-free paper

Springer is part of Springer Science+Business Media (www.springer.com)

To our ancestors and parents, as always

Preface

Kinematically redundant manipulators are those having more degrees of freedom (DOF) than required to perform a given end-effector primary task. One fundamental issue in operating such a robot system is the redundancy-resolution problem (or so-called inverse kinematics problem). An inverse-kinematics scheme is called repetitive if it maps closed paths in the task space (i.e., cyclic sequences of tasks) to closed trajectories in the configuration space (i.e., cyclic sequences of configurations). Nonrepetitive problem is that the joint angles may not return to their initial values after the end-effector traces a closed path in its workspace. Nonrepetitive problem results in a joint-angle drift phenomenon and may induce a problem that the manipulator's behavior is hard to predict; it is then less efficient to readjust the manipulator's configuration after every cycle.

In this book we present four typical motion planning schemes. One of them is online repetitive motion planning (RMP) scheme, or so-called cyclic motion generation (CMG) scheme, and its proof based on neural dynamic methods is given in detail. Then, some other optimization schemes, which can be viewed as the extensions of the RMP scheme, are developed and investigated for the purpose of repetitive motion planning. Furthermore, we unify them as quadratic programs (QPs). Moreover, some important QP solvers (including neural networks and numerical algorithms) are employed to solve the resultant QPs. Computer-simulation results based on various robotic models demonstrate the effectiveness of the proposed RMP schemes. For substantiating the physical realizability of the RMP schemes via QP formulation, one of the RMP schemes is applied to an actual planar six-DOF robot manipulator (or to say, robot arm). More specifically, we have the following organization.

In Chap. 1, we first discuss the importance of the RMP research and review the recent results on RMP. Then we present three kinds of manipulators. Their Jacobian matrices are derived in detail for further discussion, simulations, and experiments.

In Chap. 2, an optimization scheme is first presented and investigated for online RMP of redundant robot manipulators. Then, three extended schemes are developed for the purpose of RMP. These four schemes are finally reformulated and unified as QP problems with different definitions of the same coefficients.

In Chap. 3, we employ the well-known gradient-based neural network (GNN) and Zhang neural network (ZNN) approaches to analyze and prove the repetitive motion performance index.

One of state-of-the-art recurrent neural networks (RNN) is dual neural network (DNN). It can solve QP in real time. The DNN is of simple piecewise linear dynamics and has global (exponential) convergence to optimal solutions. In Chap. 4, we present such a DNN and its design method.

In addition to DNN, another type of neural network has been widely used in recent years, i.e., primal–dual neural network (PDNN). This kind of network includes mainly a traditional primal–dual neural network, a linear variational inequality (LVI) based PDNN, and a simplified LVI-based PDNN, presented in Chap. 5.

Different from the neural-network-based QP solvers, the numerical algorithms as QP solvers are very important as well, especially in the real-time kinematic resolution via today's digital computers. In Chaps. 6 and 7, we present numerical algorithms 94LVI and E47, respectively.

In Chaps. 8, 9, and 10, we present computer simulation examples based on planar multilink manipulators, PUMA560, and PA10 robots to demonstrate the effectiveness of the presented RMP scheme and the corresponding QP solvers.

To demonstrate the hardware realizability and efficacy of the QP-based methods for solving the nonrepetitive problem, Chap. 11 gives a repetitive motion planning and control (RMPC) scheme, and realizes this scheme on a physical planar six-DOF push-rod-joint (PRJ) manipulator. To control the real PRJ manipulator, this scheme considers variable joint-velocity limits and joint-limit margins. To decrease the model disturbance and computational round-off errors, this scheme also considers the position-error feedback. Then, the scheme is reformulated as a QP problem. Due to control via a digital computer, a discrete-time QP solver, termed piecewise linear equation (PLE) based numerical algorithm (i.e., numerical algorithm 94LVI), is presented to solve the QP problem. For comparison, both of the nonrepetitive and repetitive motions are performed on the six-DOF PRJ manipulator to track square, B-shaped, and circular paths. Theoretical analysis and experimental results validate the physical realizability and effectiveness of the RMPC scheme. Position-error analysis further verifies the accuracy of this scheme.

In summary, the book solves the nonrepetitive motion problem, which has stood in the areas of robotics and control for 30 years (specifically, since the work of Klein and Huang in 1983). The QP technique is exploited in this RMPC research with rich verification of simulations and experiments, while the traditional methods relate to pseudoinverse-type solutions. Now, the door to the industrial applications of redundant manipulators is completely open, as the difficult nonrepetitive motion problem has been solved truly, systematically, and methodologically. Without doubt, this book can be extended. Any comments or suggestions are welcome. The authors can be contacted via e-mails zhynong@mail.sysu.

edu.cn and drzhangzhijun@gmail.com. The web page of Yunong Zhang is http://sist.sysu.edu.cn/~zhynong/.

Guangzhou, China
February 2013

Yunong Zhang
Zhijun Zhang

Acknowledgements

During the work on this book, we have had the pleasure of discussing its various aspects and results with many cooperators and students. We highly appreciate their contributions, which particularly allowed us to improve our book manuscript. Especially valuable help was provided by Dongsheng Guo, Huarong Wu, Lin Xiao, Bingguo Mu, Xiaotian Yu, Jun Li, Senbo Fu, Jinhao Chen, Yonghua Yin, and Weibing Li. We are grateful to them for their help and suggestions.

The continuous support of our research by the National Science Foundation of China under Grants 61075121 and 60935001 and by the Fundamental Research Funds for the Central Universities of China is gratefully acknowledged here either.

Besides, we would like to thank sincerely the editors for their very important and constructive comments and suggestions provided, in addition to their time and efforts spent in handling this book.

We are always very grateful to the nice people (especially the staff in Springer) for their strong support and encouragement during the preparation and publishing of this book.

Contents

Acronyms

RMP	Repetitive motion planning
DOF	Degrees of freedom
DNN	Dual neural network
QP	Quadratic program or Quadratic programming
LVI	Linear variational inequality
PDNN	Primal–dual neural network
KKT	Karush–Kuhn–Tucker
LP	Linear program or Linear programming
LVI-PDNN	LVI-based primal–dual neural network
RNN	Recurrent neural network
GNN	Gradient neural network or Gradient-based neural network
ZNN	Zhang neural network
PLE	Piecewise linear equations
PRJ	Push-rod-joint

Chapter 1
Fundamentals

Abstract In this chapter we first discuss the importance of the research of the repetitive motion planning (RMP) of redundant robots and review the recent results on RMP. Then we present three kinds of robotics. Especially, the Jacobian matrices of a five-link planar robot manipulator, the PUMA560 manipulator, and the PA10 manipulator are derived in detail for further discussion, simulations, and experiments.

1.1 Introduction

Kinematically redundant manipulators are those having more degrees of freedom (DOF) than required for end-effectors' position and orientation in the kinematic sense [1]. Redundant degrees of freedom (DOF) make them possible to achieve simultaneously many useful objectives, such as collision avoidance with obstacles in the operational space, joint limits avoidance, and/or singular configurations avoidance when the manipulator moves [2].

1.1.1 Redundancy of the Robots

As for the redundant manipulators, since they have more DOF, multiple solutions exist. This redundancy (or to say, multiplicity) of inverse-kinematics mappings has complicated the issue of manipulators' motion planning and control substantially, in addition to the complication from the mappings' nonlinearity. Given readily the workspace path of the end-effector (at least at that time instant t), the online computation of corresponding feasible joint trajectories for the redundant manipulator is referred to as redundancy resolution [3].

To take full advantage of the redundancy, various computational schemes have been developed. Conventionally, the general resolution of redundancy problem is obtained in the form of pseudoinverse-type solution. It includes a minimum-norm particular solution and a homogeneous solution [4, 5]. Most researchers have applied this pseudoinverse-type formulation to resolving the redundancy problem by considering different optimization criteria, such as minimum joint-velocity norm and task priority control. However, among those techniques, the physical limits

Y. Zhang, Z. Zhang, *Repetitive Motion Planning and Control of Redundant Robot Manipulators*, DOI 10.1007/978-3-642-37518-7_1,

(e.g., joint limits and joint-velocity limits) are usually not (or difficult to be) taken into account, as these inequality constraints are relatively not easy to handle via pseudoinverse-type formulation [6]. However, if these physical limits are not considered, then the resolved solution may saturate or be inapplicable. As a result, the tracking error increases appreciably, not to mention physical damages possibly caused by a commanded joint hitting its physical bound.

1.1.2 Neural Networks for Resolving the Redundancy

In recent years, among parallel distributed computational models, recurrent neural networks (RNNs) have been presented and applied to the redundancy resolution of robot manipulators, e.g., [7–11]. In particular, two neural networks, namely the pseudoinverse neural network [10] and the linear-programming neural network [11], were applied to the minimum infinity-norm kinematic control in [7, 12]. The solution achieved therein explicitly minimizes the joint-velocity magnitude. To improve network-structure and computational-efficiency, in this book, four efficient neural networks are presented. The first neural network is dual neural network (DNN). It can solve QP in real time. The dual neural network is of simple piecewise linear dynamics and has global (exponential) convergence to optimal solutions. Based on the duality theory [13], a special kind of recurrent neural network (namely, primal–dual neural network, PDNN) can also be developed and exploited for kinematic control of redundant manipulators. The third neural network is the LVI-based primal–dual neural network, which is designed based on the QP-LVI conversion and Karush–Kuhn–Tucker (KKT) conditions. With simple piecewise linear dynamics and global (exponential) convergence to optimal solutions, it can handle general QP and LP (linear-programming) problems in the inverse-free manner. To further reduce the implementation complexity and also motivated by the authors' design experience, a simplified LVI-based primal–dual neural network (simplified LVI-PDNN) is presented.

1.1.3 Numerical Algorithms for Resolving the Redundancy

For the joint-constrained inverse kinematics, numerical algorithms of redundancy resolution have also been developed. For example, the Jacobian matrix was augmented in [14] by incorporating joint and velocity limits, and it required a routine predicting which constraints might be violated. The joint and velocity limits were also taken into consideration in [15] by applying the Gram–Schmidt orthogonalization procedure. Cheng et al. [6] formulated the constrained kinematic redundancy problem into a QP form. A compact QP method, using Gaussian elimination with partial pivoting, was finally developed [16]. In this book, two important and efficient numerical algorithms are developed, which are based on the linear variational inequality reformulation. One of them is called 94LVI algorithm, the other is E47 algorithm. It is worth mentioning that these two algorithms are readily performed on digital computer to control the physical redundant robot manipulators.

1.1.4 Nonrepetitive Motion Problem

When the end-effector traces a closed path in its workspace, the joint angles may not return to their initial values after completing such an end-effector task. In other words, the trajectories obtained in the joint space may not be closed ones. This is called the joint-angle drift phenomenon or nonrepetitive motion problem. This may induce a problem that the manipulator's behavior is hard to predict, and it is less efficient to readjust the manipulator's configuration via self-motion. Klein and Huang [17] were the first to observe this nonrepetitive motion phenomenon for the case of the pseudoinverse control of a planar manipulator.

In this book we present a typical repetitive motion planning (RMP) scheme (i.e., joint-angle drift remedy scheme), give its proof, which is based on neural dynamic methods, and unify the scheme as a QP. In addition, some important QP solvers (including neural networks and numerical algorithms) are employed to solve the resultant QP. Furthermore, computer simulation results based on various robotic models demonstrate the effectiveness of the proposed RMP scheme. For substantiating the physical realizability of the RMP scheme based on QP, one variant of the RMP scheme is applied to a physical planar six-DOF robot manipulator.

1.2 Forward Kinematic Equation

The end-effector position-and-orientation vector $r(t) \in R^m$ in Cartesian space and the joint-space vector $\theta(t) \in R^n$ are related nonlinearly by the forward-kinematic equation $r(t) = f(\theta(t))$, which is abbreviated as

$$r = f(\theta). \tag{1.1}$$

The above equation is, however, usually difficult to solve inversely due to its nonlinearity and redundancy. The problem is then considered at the joint-velocity level:

$$J(\theta)\dot{\theta} = \dot{r}, \tag{1.2}$$

where $\dot{r}$ and $\dot{\theta}$ denote the m-dimensional end-effector velocity vector and the n-dimensional joint-velocity vector, respectively. The Jacobian matrix $J(\theta)$ is defined as $J(\theta) = \partial f(\theta)/\partial\theta \in R^{m\times n}$. For redundant manipulators, because $m < n$, Eqs. (1.1) and (1.2) are under-determined and thus admit an infinite number of solutions.

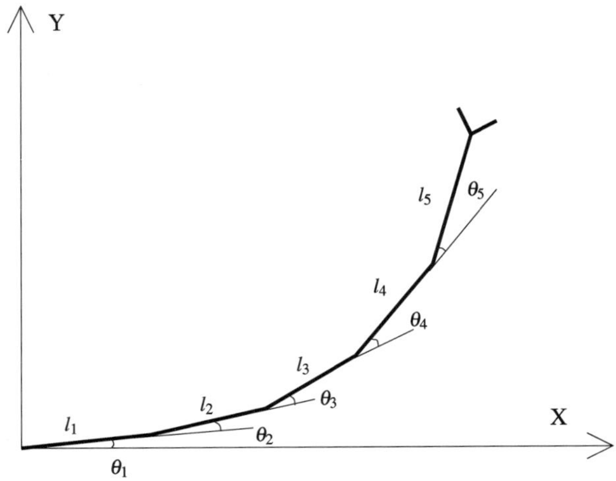

Fig. 1.1 Geometry of five-link planar manipulator

The conventional pseudoinverse-type solution to (1.2) is generally given as

$$\dot{\theta} = J^{\dagger}\dot{r} + \big(I - J^{\dagger}J\big)\tilde{z}, \tag{1.3}$$

where $J^{\dagger} \in R^{n\times m}$ denotes the pseudoinverse of J, and $\tilde{z} \in R^{n}$ is an arbitrary vector usually selected by using some optimization criteria, e.g., those for singularity avoidance, obstacle avoidance, and/or task priority control.

1.3 Geometry of Five-Link Planar Robot Manipulator

We show in Fig. 1.1 the geometry of the five-link planar robot manipulator [18]. The manipulator has five planar revolute joints, i.e., $n = 5$. In terms of joint angles θ_i $(i = 1, 2, \ldots, 5)$ defined in Fig. 1.1, we can readily get the forward-kinematic equation:

$$r := \begin{bmatrix} r_X \\ r_Y \end{bmatrix} = \begin{bmatrix} l_1c_1 + l_2c_2 + l_3c_3 + l_4c_4 + l_5c_5 \\ l_1s_1 + l_2s_2 + l_3s_3 + l_4s_4 + l_5s_5 \end{bmatrix} = f(\theta),$$

where $r \in R^2$ (i.e., $m = 2$), $\theta := [\theta_1, \theta_2, \ldots, \theta_5]^{\mathrm{T}} \in R^5$, with superscript ^{T} denoting the transpose of a vector or matrix, and with the following definitions of variables:

$$\begin{aligned}
c_1 &:= \cos(\theta_1), & s_1 &:= \sin(\theta_1),\\
c_2 &:= \cos(\theta_1 + \theta_2), & s_2 &:= \sin(\theta_1 + \theta_2),\\
c_3 &:= \cos\big(\textstyle\sum_{i=1}^{3} \theta_i\big), & s_3 &:= \sin\big(\textstyle\sum_{i=1}^{3} \theta_i\big),\\
c_4 &:= \cos\big(\textstyle\sum_{i=1}^{4} \theta_i\big), & s_4 &:= \sin\big(\textstyle\sum_{i=1}^{4} \theta_i\big),\\
c_5 &:= \cos\big(\textstyle\sum_{i=1}^{5} \theta_i\big), & s_5 &:= \sin\big(\textstyle\sum_{i=1}^{5} \theta_i\big).
\end{aligned}$$

Given a path $r(t) \in R^2$, solving directly $f(\cdot)$ for $\theta(t) \in R^5$ is difficult in view of the nonlinearity and redundancy of $f(\cdot)$, together with the limited physical ranges of θ and $\dot{\theta}$. Note that, about this five-link manipulator, for example, the physical upper/lower limits of θ and $\dot{\theta}$ are respectively $\pm[3\pi/4, 3\pi/4, 3\pi/4, 3\pi/4, 3\pi/4]^{\mathrm{T}}$ rad and $\pm[\pi/3, \pi/3, \pi/3, \pi/3, \pi/3]^{\mathrm{T}}$ rad/s.

To solve the redundancy resolution problem at the joint-velocity level [i.e., (1.2)], we have to obtain the Jacobian matrix of $f(\cdot)$ by differentiating (1.1), which is shown as

$$J = \begin{bmatrix} J_{11} & J_{12} & J_{13} & J_{14} & J_{15} \\ J_{21} & J_{22} & J_{23} & J_{24} & J_{25} \end{bmatrix}, \tag{1.4}$$

where

$$\begin{aligned}
J_{11} &= -l_1 s_1 - l_2 s_2 - l_3 s_3 - l_4 s_4 - l_5 s_5,\\
J_{12} &= -l_2 s_2 - l_3 s_3 - l_4 s_4 - l_5 s_5,\\
J_{13} &= -l_3 s_3 - l_4 s_4 - l_5 s_5,\\
J_{14} &= -l_4 s_4 - l_5 s_5,\\
J_{15} &= -l_5 s_5,\\
J_{21} &= l_1 c_1 + l_2 c_2 + l_3 c_3 + l_4 c_4 + l_5 c_5,\\
J_{22} &= l_2 c_2 + l_3 c_3 + l_4 c_4 + l_5 c_5,\\
J_{23} &= l_3 c_3 + l_4 c_4 + l_5 c_5,\\
J_{24} &= l_4 c_4 + l_5 c_5,\\
J_{25} &= l_5 c_5.
\end{aligned}$$

In the computer simulations presented in Chap. 8, we set $l_1 = l_2 = l_3 = l_4 = l_5 =$ 1 m and apply effectively the neural-network approaches to this five-link planar robot manipulator. Both joint-physical-limits avoidance and joint-angle-drift elimination are achieved successfully, with high computational precision and efficiency demonstrated.

1.4 Geometry of PA10

The PA10 manipulator has seven DOF (four rotation axes and three pivot axes) as illustrated in Fig. 1.2, and the corresponding parameters of the joints are shown in Table 1.1 [19, 20]. The key specifications of the PA10 manipulator are that the full length of the manipulator is 1.345 m, the weight of the manipulator is 35 kgf (i.e., 343 N), the maximum combined speed with all axes is 1.55 m/s, the payload weight

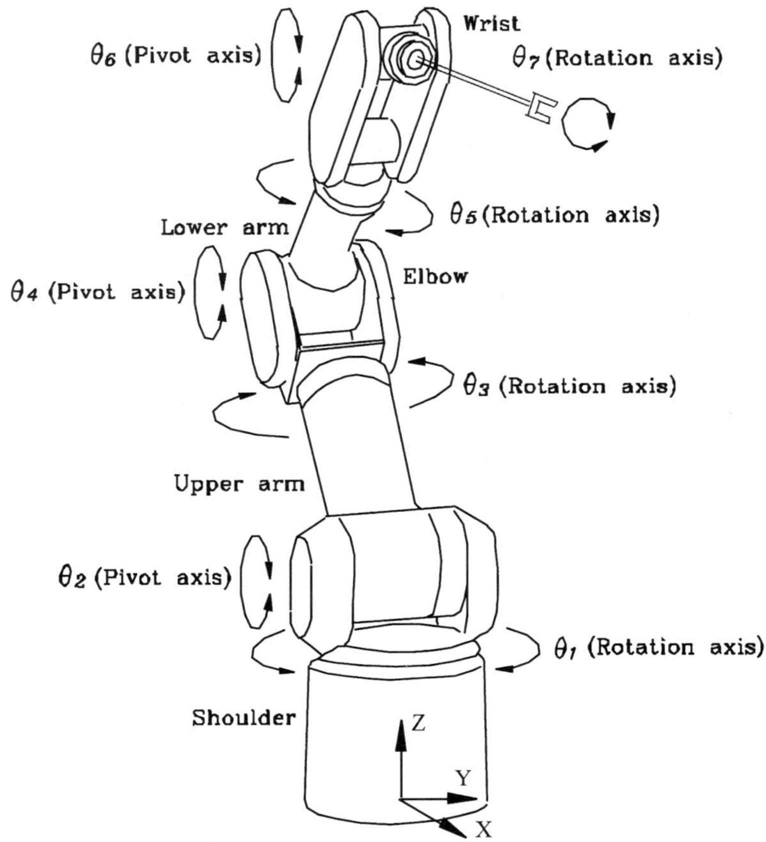

Fig. 1.2 Mechanical configuration of PA10 robot arm

Table 1.1 Joint limits of PA10 manipulator

#	Joint	axis	$\theta^{\pm}$
1	Shoulder 1	Rotating	$\pm\pi$ rad
2	Shoulder 2	Pivoting	± 1.7637 rad
3	Shoulder 3	Rotating	$\pm\pi$ rad
4	Elbow 1	Pivoting	± 2.6831 rad
5	Elbow 2	Rotating	$\pm 3\pi/2$ rad
6	Wrist 1	Pivoting	$\pm\pi$ rad
7	Wrist 2	Rotating	$\pm 2\pi$ rad

is 10 kgf (i.e., 98 N), and the output torque is 9.8 N m. The homogeneous transformation matrix T_{07}, which specifies the position and orientation of the end-effector with respect to the base coordinate system, is the chain product of successive coordinate transformation matrices of $T_{i-1,i}$ for $i = 1, 2, \ldots, 7$ expressed as

$$T_{07} = T_{01}T_{12}T_{23}T_{34}T_{45}T_{56}T_{67} = \begin{bmatrix} \mathbf{n} & \mathbf{s} & \mathbf{a} & \mathbf{p} \\ 0 & 0 & 0 & 1 \end{bmatrix}, \tag{1.5}$$

where $\mathbf{n} \in R^3$ is the normal vector of the end-effector in the base coordinate system, $\mathbf{s} \in R^3$ is the sliding vector of the end-effector, $\mathbf{a} \in R^3$ is the approach vector of the end-effector, and $\mathbf{p} = [x, y, z]^{\mathrm{T}}$ is the position vector of the end-effector. Besides,

$$T_{01} = \begin{bmatrix} c_1 & -s_1 & 0 & 0 \\ s_1 & c_1 & 0 & 0 \\ 0 & 0 & 1 & 0 \\ 0 & 0 & 0 & 1 \end{bmatrix}, \qquad T_{12} = \begin{bmatrix} c_2 & -s_2 & 0 & 0 \\ 0 & 0 & 1 & 0 \\ -s_2 & -c_2 & 0 & 0 \\ 0 & 0 & 0 & 1 \end{bmatrix},$$

$$T_{23} = \begin{bmatrix} c_3 & -s_3 & 0 & 0 \\ 0 & 0 & -1 & -d_3 \\ s_3 & c_3 & 0 & 0 \\ 0 & 0 & 0 & 1 \end{bmatrix}, \qquad T_{34} = \begin{bmatrix} c_4 & -s_4 & 0 & 0 \\ 0 & 0 & 1 & 0 \\ -s_4 & -c_4 & 0 & 0 \\ 0 & 0 & 0 & 1 \end{bmatrix},$$

$$T_{45} = \begin{bmatrix} c_5 & -s_5 & 0 & 0 \\ 0 & 0 & -1 & -d_5 \\ s_5 & c_5 & 0 & 0 \\ 0 & 0 & 0 & 1 \end{bmatrix}, \qquad T_{56} = \begin{bmatrix} c_6 & -s_6 & 0 & 0 \\ 0 & 0 & 1 & 0 \\ -s_6 & -c_6 & 0 & 0 \\ 0 & 0 & 0 & 1 \end{bmatrix},$$

$$T_{67} = \begin{bmatrix} c_7 & -s_7 & 0 & 0 \\ 0 & 0 & -1 & -d_7 \\ s_7 & c_7 & 0 & 0 \\ 0 & 0 & 0 & 1 \end{bmatrix}.$$

$c_i = \cos\theta_i$, and $s_i = \sin\theta_i$. The lengths of the arms are $d_3 = 0.450$ m, $d_5 = 0.5$ m, and $d_7 = 0.080$ m. The Jacobian of the PA10 manipulator can be derived as follows:

$$J = \begin{bmatrix} J_{11} & J_{12} & J_{13} & J_{14} & J_{15} & J_{16} & J_{17} \\ J_{21} & J_{22} & J_{23} & J_{24} & J_{25} & J_{26} & J_{27} \\ J_{31} & J_{32} & J_{33} & J_{34} & J_{35} & J_{36} & J_{37} \\ J_{41} & J_{42} & J_{43} & J_{44} & J_{45} & J_{46} & J_{47} \\ J_{51} & J_{52} & J_{53} & J_{54} & J_{55} & J_{56} & J_{57} \\ J_{61} & J_{62} & J_{63} & J_{64} & J_{65} & J_{66} & J_{67} \end{bmatrix},$$

where

$$J_{11} = -s_1 q_{114} - c_1 q_{234}, \qquad J_{12} = c_1 p_{23}, \qquad J_{13} = s_1 s_2 p_{33} - c_2 p_{32},$$

$$J_{14} = q_{423} p_{43} - q_{433} p_{42}, \qquad J_{15} = -q_{422} p_{53} + q_{432} p_{52},$$

$$J_{16} = q_{522} p_{63} - q_{532} p_{62}, \qquad J_{17} = 0,$$

$$J_{21} = c_1 q_{114} - s_1 q_{234}, \qquad J_{22} = s_1 p_{23},$$

$$J_{23} = c_2 p_{31} - c_1 s_2 p_{33}, \qquad J_{24} = q_{433} p_{41} - q_{413} p_{43},$$

$$J_{25} = -q_{432}p_{51} + q_{412}p_{53}, \qquad J_{26} = q_{532}p_{61} - q_{512}p_{63}, \qquad J_{27} = 0,$$
$$J_{31} = 0, \qquad J_{32} = -s_1 p_{22} - c_1 p_{21}, \qquad J_{33} = c_1 s_2 p_{32} - s_1 s_2 p_{31},$$
$$J_{34} = q_{413}p_{42} - q_{423}p_{41}, \qquad J_{35} = -q_{412}p_{52} + q_{422}p_{51},$$
$$J_{36} = q_{512}p_{62} - q_{522}p_{61}, \qquad J_{37} = 0,$$
$$J_{41} = 0, \qquad J_{42} = -s_1, \qquad J_{43} = c_1 s_2, \qquad J_{44} = q_{413},$$
$$J_{45} = -q_{412}, \qquad J_{46} = q_{512}, \qquad J_{47} = -q_{612},$$
$$J_{51} = 0, \qquad J_{52} = c_1, \qquad J_{53} = s_1 s_2, \qquad J_{54} = q_{423},$$
$$J_{55} = q_{422}, \qquad J_{56} = q_{522}, \qquad J_{57} = -q_{622},$$
$$J_{61} = 1, \qquad J_{62} = 0, \qquad J_{63} = c_2, \qquad J_{64} = q_{433},$$
$$J_{65} = -q_{432}, \qquad J_{66} = q_{532}, \qquad J_{67} = -q_{632}$$

with

$$q_{414} = c_5 s_6 d_7, \qquad q_{424} = -c_6 d_7 - d_5, \qquad q_{434} = s_5 s_6 d_7,$$
$$q_{314} = c_4 q_{414} - s_4 q_{424}, \qquad q_{324} = q_{434},$$
$$q_{334} = -s_4 q_{414} - c_4 q_{424}, \qquad q_{214} = c_3 q_{314} - s_3 q_{434},$$
$$q_{224} = -q_{334} - d_3, \qquad q_{234} = s_3 q_{314} + c_3 q_{434},$$
$$q_{114} = c_2 q_{214} - s_2 q_{224}, \qquad q_{134} = -s_2 q_{214} - c_2 q_{224},$$
$$q_{411} = (c_1 c_2 c_3 - s_1 s_3) c_4 - c_1 s_2 s_4,$$
$$q_{412} = -(c_1 c_2 c_3 - s_1 s_3) s_4 - c_1 s_2 c_4,$$
$$q_{413} = -c_1 c_2 c_3 - s_1 c_3, \qquad q_{421} = (s_1 c_2 c_3 + c_1 s_3) c_4 - s_1 s_2 s_4,$$
$$q_{422} = -(s_1 c_2 c_3 + c_1 s_3) s_4 - s_1 s_2 c_4,$$
$$q_{423} = -s_1 c_2 s_3 + c_1 c_3, \qquad q_{431} = -s_2 c_3 c_4 - c_2 s_4,$$
$$q_{432} = s_2 c_3 s_4 - c_2 c_4, \qquad q_{433} = s_2 s_3,$$
$$q_{511} = q_{411} c_5 + q_{413} s_5, \qquad q_{512} = -q_{411} s_5 + q_{413} c_5,$$
$$q_{513} = -q_{412}, \qquad q_{521} = q_{421} c_5 + q_{423} s_5,$$
$$q_{522} = -q_{421} s_5 + q_{423} c_5, \qquad q_{523} = -q_{422},$$
$$q_{531} = q_{431} c_5 + q_{433} s_5, \qquad q_{532} = -q_{431} s_5 + q_{433} c_5,$$
$$q_{533} = -q_{432}, \qquad q_{612} = -q_{511} s_6 + q_{412} c_6,$$
$$q_{622} = -q_{521} s_6 + q_{422} c_6, \qquad q_{632} = -q_{531} s_6 + q_{432} c_6,$$
$$p_{21} = c_1 c_2 q_{214} - c_1 s_2 q_{224} - s_1 q_{234},$$
$$p_{22} = s_1 c_2 q_{214} - c_1 s_2 q_{224} + c_1 q_{234},$$

$$p_{23} = -s_2 q_{214} - c_2 q_{224},$$

$$p_{31} = (c_1c_2c_3 - s_1s_3)q_{314} - (c_1c_2c_3 + s_1c_3)q_{434} + c_1s_2q_{334},$$

$$p_{32} = (s_1c_2c_3 + c_1s_3)q_{314} + (-s_1c_2s_3 + c_1c_3)q_{434} + s_1s_2q_{334},$$

$$p_{33} = -s_2c_3q_{314} + s_2s_3q_{434} + c_2q_{334},$$

$$p_{41} = q_{411}q_{414} + q_{412}q_{424} + q_{413}q_{434},$$

$$p_{42} = q_{421}q_{414} + q_{422}q_{424} + q_{423}q_{434},$$

$$p_{43} = q_{431}q_{414} + q_{432}q_{424} + q_{433}q_{434},$$

$$p_{51} = q_{511}s_6d_7 + q_{513}c_6d_7,$$

$$p_{52} = q_{521}s_6d_7 + q_{523}c_6d_7,$$

$$p_{53} = q_{531}s_6d_7 + q_{533}c_6d_7, \qquad p_{61} = -q_{612}d_7,$$

$$p_{62} = -q_{622}d_7, \quad \text{and} \quad p_{63} = -q_{632}d_7.$$

1.5 Geometry of PUMA560

The geometry of PUMA560 is shown in Fig. 1.3, which has six joints. In general, the PUMA560 manipulator is nonredundant when position and orientation are both taken into consideration. However, when we only consider the position of the end-point of an attached tool at the end-effector, it becomes a functionally redundant robot [21].

The related parameters of PUMA560 are shown in Table 1.2. With the above parameters (i.e., $a_2 = 0.4318$ m, $a_3 = 0.0203$ m, $d_3 = 0.4318$ m, $d_4 = 0.4318$ m, $d_6 = 0.25625$ m), we have the following matrices:

$$T_{01} = \begin{bmatrix} c_1 & 0 & s_1 & 0 \\ s_1 & 0 & -c_1 & 0 \\ 0 & 1 & 0 & 0 \\ 0 & 0 & 0 & 1 \end{bmatrix}, \qquad T_{12} = \begin{bmatrix} c_2 & -s_2 & 0 & a_2c_2 \\ s_2 & c_2 & 0 & a_2s_2 \\ 0 & 0 & 1 & 0 \\ 0 & 0 & 0 & 1 \end{bmatrix},$$

$$T_{23} = \begin{bmatrix} c_3 & 0 & -s_3 & a_3c_3 \\ s_3 & 0 & c_3 & a_3s_3 \\ 0 & -1 & 0 & d_3 \\ 0 & 0 & 0 & 1 \end{bmatrix}, \qquad T_{34} = \begin{bmatrix} c_4 & 0 & s_4 & 0 \\ s_4 & 0 & -c_4 & 0 \\ 0 & 1 & 0 & d_4 \\ 0 & 0 & 0 & 1 \end{bmatrix},$$

$$T_{45} = \begin{bmatrix} c_5 & 0 & -s_5 & 0 \\ s_5 & 0 & c_5 & 0 \\ 0 & -1 & 0 & 0 \\ 0 & 0 & 0 & 1 \end{bmatrix}, \qquad T_{56} = \begin{bmatrix} c_6 & 0 & -s_6 & 0 \\ s_6 & c_6 & 0 & 0 \\ 0 & 0 & 1 & d_6 \\ 0 & 0 & 0 & 1 \end{bmatrix},$$

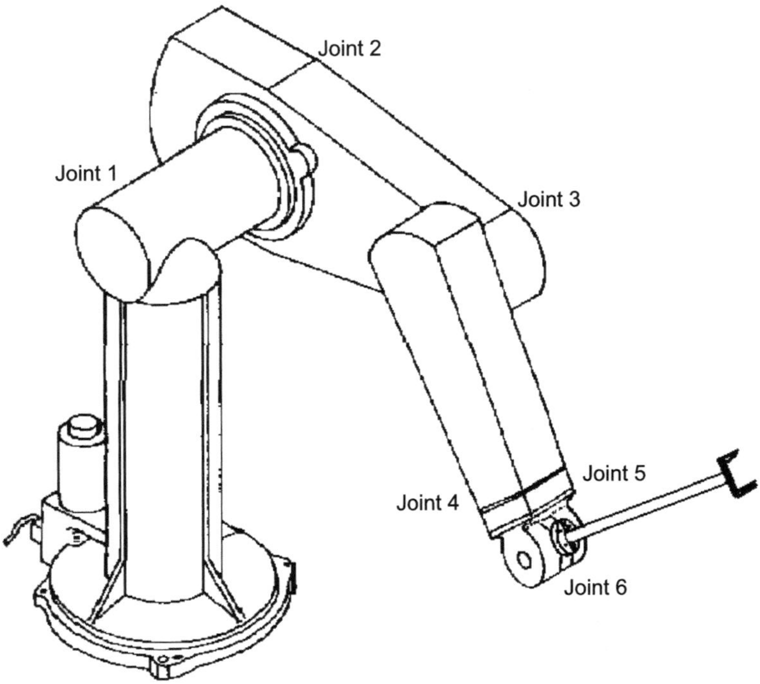

Fig. 1.3 Mechanical configuration of PUMA560 manipulator

Table 1.2 Parameters of PUMA560 manipulator joints

Joint	θ_i	$\hat{\alpha}_i$	a_i	d_i
1	θ_1	90° ($\pi/2$ rad)	0 m	0 m
2	θ_2	0° (0 rad)	0.4318 m	0 m
3	θ_3	−90° ($-\pi/2$ rad)	0.0203 m	0.4318 m
4	θ_4	90° ($\pi/2$ rad)	0 m	0.4318 m
5	θ_5	−90° ($-\pi/2$ rad)	0 m	0 m
6	θ_6	0° (0 rad)	0 m	0.25625 m

where $c_i = \cos\theta_i$, and $s_i = \sin\theta_i$ with $i = 1, 2, 3, 4, 5, 6$. Then, from the above matrices we have

$$DT_{01} = \begin{bmatrix} -s_1 & 0 & c_1 & 0 \\ c_1 & 0 & s_1 & 0 \\ 0 & 0 & 0 & 0 \\ 0 & 0 & 0 & 0 \end{bmatrix}, \qquad DT_{12} = \begin{bmatrix} -s_2 & -c_2 & 0 & -a_2 s_2 \\ c_2 & -s_2 & 0 & a_2 c_2 \\ 0 & 0 & 0 & 0 \\ 0 & 0 & 0 & 0 \end{bmatrix},$$

$$DT_{23} = \begin{bmatrix} -s_3 & 0 & -c_3 & -a_3 s_3 \\ c_3 & 0 & -s_3 & a_3 c_3 \\ 0 & 0 & 0 & 0 \\ 0 & 0 & 0 & 0 \end{bmatrix}, \qquad DT_{34} = \begin{bmatrix} -s_4 & 0 & c_4 & 0 \\ c_4 & 0 & -s_4 & 0 \\ 0 & 0 & 0 & 0 \\ 0 & 0 & 0 & 0 \end{bmatrix},$$

$$DT_{45} = \begin{bmatrix} -s_5 & 0 & -c_5 & 0 \\ c_5 & 0 & -s_5 & 0 \\ 0 & 0 & 0 & 0 \\ 0 & 0 & 0 & 0 \end{bmatrix}, \qquad DT_{56} = \begin{bmatrix} -s_6 & 0 & -c_6 & 0 \\ c_6 & -s_6 & 0 & 0 \\ 0 & 0 & 0 & 0 \\ 0 & 0 & 0 & 0 \end{bmatrix},$$

and

$$T^{6\text{v}} = T_{56}[0.0;\,0.0;\,0.0;\,1.0], \qquad T^{56\text{v}} = T_{45}T^{6\text{v}},$$
$$T^{456\text{v}} = T_{34}T^{56\text{v}}, \qquad T^{3456\text{v}} = T_{23}T^{456\text{v}},$$
$$T^{23456\text{v}} = T_{12}T^{3456\text{v}}, \qquad J_1 = DT_{01}T^{23456\text{v}},$$
$$D_2T^{3456\text{v}} = DT_{12}T^{3456\text{v}}, \qquad J_2 = T_{01}D_2T^{3456\text{v}},$$
$$D_3T^{456\text{v}} = DT_{23}T^{456\text{v}}, \qquad T^2D_3T^{456\text{v}} = T_{12}D_3T^{456\text{v}},$$
$$J_3 = T_{01}T^2D_3T^{456\text{v}}, \qquad D_4T^{56\text{v}} = DA_{34}T^{56\text{v}},$$
$$T^3D_4T^{56\text{v}} = T_{23}D_4T^{56\text{v}}, \qquad T^{23}D_4T^{56\text{v}} = T_{12}T^3D_4T^{56\text{v}},$$
$$J_4 = T_{01}T^{23}D_4A^{56\text{v}}, \qquad D_5T^{6\text{v}} = DT_{45}T^{6\text{v}},$$
$$T^4D_5T^{6\text{v}} = T_{34}D_5T^{6\text{v}}, \qquad T^{34}D_5T^{6\text{v}} = T_{23}T^4D_5T^{6\text{v}},$$
$$T^{234}D_5T^{6\text{v}} = T_{12}T^{34}D_5T^{6\text{v}}, \qquad J_5 = T_{01}T^{234}D_5T^{6\text{v}},$$
$$D_{6\text{v}} = DA_{56}[0.0;\,0.0;\,0.0;\,1.0], \qquad T^5D_{6\text{v}} = T_{45}D_{6\text{v}},$$
$$T^{45}D_{6\text{v}} = T_{34}T^5D_{6\text{v}}, \qquad T^{345}D_{6\text{v}} = T_{23}T^{45}D_{6\text{v}},$$
$$T^{2345}D_{6\text{v}} = T_{12}T^{345}D_{6\text{v}}, \qquad J_6 = T_{01}T^{2345}D_{6\text{v}}.$$

Thus,

$$J(:,1) = J_1(1:3,1), \qquad J(:,2) = J_2(1:3,1),$$
$$J(:,3) = J_3(1:3,1), \qquad J(:,4) = J_4(1:3,1),$$
$$J(:,5) = J_5(1:3,1), \qquad J(:,6) = J_6(1:3,1).$$

Note that, in the above derivations of the geometry of PUMA560, MATLAB notation is used. For example, $[0.0;\,0.0;\,0.0;\,1.0] = [0.0, 0.0, 0.0, 1.0]^{\mathrm{T}}$, and $J(:,i)$ denotes the ith column of a matrix J with $i = 1, 2, 3, 4, 5, 6$.

1.6 Chapter Conclusions

In this chapter, we discussed the importance of the research of the repetitive motion planning (RMP) and reviewed the recent results on RMP. Additionally, the Jacobian matrices of the five-link planar robot manipulator, the PUMA560 manipulator, and the PA10 manipulator have been derived as a basis of the book.

References

1. Sciavicco L, Siciliano B (2000) Modeling and control of robot manipulators. Springer, London
2. Galicki M (2006) Path-constrained control of a redundant manipulator in a task space. Robot Auton Syst 54(3):234–243
3. Conkur ES (2003) Path following algorithm for highly redundant manipulators. Robot Auton Syst 45(1):1–22
4. Khatib O, Bowling A (1996) Optimization of the inertial and acceleration characteristics of manipulators. In: Proceedings of IEEE international conference on robotics and automation, pp 2883–2889
5. Liegeois A (1977) Automatic supervisory control of the configuration and behavior of multibody mechanisms. IEEE Trans Syst Man Cybern 7(12):868–871
6. Cheng FT, Chen TH, Sun YY (1994) Resolving manipulator redundancy under inequality constraints. IEEE Trans Robot Autom 10(1):65–71
7. Ding H, Wang J (1999) Recurrent neural networks for minimum infinity-norm kinematic control of redundant manipulators. IEEE Trans Syst Man Cybern, Part A, Syst Hum 29(3):269–276
8. Mao Z, Hsia TC (1997) Obstacle avoidance inverse kinematics solution of redundant robots by neural networks. Robotica 15(1):3–10
9. Tang WS, Wang J (2001) A recurrent neural network for minimum infinity-norm kinematic control of redundant manipulators with an improved problem formulation and reduced architectural complexity. IEEE Trans Syst Man Cybern, Part B, Cybern 31(1):98–105
10. Wang J (1997) Recurrent neural networks for computing pseudoinverses of rank-deficient matrices. SIAM J Sci Comput 18(5):1479–1493
11. Xia Y (1996) A new neural network for solving linear programming problems and its application. IEEE Trans Neural Netw 7(2):525–529
12. Zhang Y (2006) Inverse-free computation for infinity-norm torque minimization of robot manipulators. Mechatronics 16(3–4):177–184
13. Zhang Y (2007) Dual neural networks: design, analysis, and application to redundant robotics. In: Kang GB (ed) Progress in neurocomputing research. Nova Science, New York, pp 41–81
14. Vukobratovic M, Kircanski M (1984) A dynamic approach to nominal trajectory synthesis for redundant manipulator. IEEE Trans Syst Man Cybern 14(4):580–586
15. Abdel-Rahman TM (1991) A generalized practical method for analytic solution of the constrained inverse kinematics problem of redundant manipulators. Int J Robot Res 10(4):382–395
16. Cheng FT, Sheu RJ, Chen TH (1995) The improved compact QP method for resolving manipulator redundancy. IEEE Trans Syst Man Cybern 25(11):1521–1530
17. Klein CA, Huang CH (1983) Review of pseudoinverse control for use with kinematically redundant manipulators. IEEE Trans Syst Man Cybern 13(3):245–250
18. Zhang Z, Tan Z, Yang Z, Lv X (2008) A dual neural network applied to drift-free resolution of five-link planar robot arm. In: Proceedings of IEEE international conference on information and automation, pp 1274–1279

19. Wang J, Zhang Y (2004) Recurrent neural networks for real-time computation of inverse kinematics of redundant manipulators. In: Sincak P, Vascak J, Hirota K (eds) Machine intelligence: quo vadis? World Scientific, Singapore, pp 299–319
20. Wang J, Hu Q, Jiang D (1999) A Lagrangian network for kinematic control of redundant robot manipulators. IEEE Trans Neural Netw 10(5):1123–1132
21. Zhang Y, Wang J (2002) A dual neural network for constrained torque optimization of kinematically redundant manipulators. IEEE Trans Syst Man Cybern, Part B, Cybern 32(5):654–662

Part I
Schemes

Chapter 2
Robotic RMP Schemes and QP Formulations

Abstract Differing from the conventional pseudoinverse-type scheme, an optimization scheme (specifically, a minimization scheme) is presented and investigated in this chapter for online RMP of redundant robot manipulators. Such an RMP scheme, which takes into account the avoidance of joint physical limits (e.g., joint-angle limits and joint-velocity limits), aims at remedying the so-called joint-angle drift problem. Then, some other optimization schemes, which can be viewed as the extensions of the RMP scheme, are developed and investigated for the purpose of repetitive motion planning. These schemes are finally reformulated and unified as QP problems with different definitions of the same coefficients.

2.1 Introduction

A robot manipulator is redundant when more DOF are observed in relation to the minimum number of DOF required to perform a given end-effector primary task [1, 2]. One fundamental issue in operating such redundant robot systems is the online redundancy-resolution problem (or to say, the inverse kinematics problem) [1–6]. Note that there are multiple feasible solutions to the redundancy-resolution problem. By properly resolving the redundancy, the robots can avoid joint physical limits and environmental obstacles, apart from optimizing various secondary criteria [1–6]. To take full advantage of the redundancy, various computational schemes, including approaches based on QP, have thus been proposed, developed, and investigated. The conventional solution to such a redundancy-resolution problem mostly takes the pseudoinverse-type formulation (i.e., a minimum-norm particular solution plus a homogeneous solution) [2, 7–10]. Recent research [1, 3–6, 11–17] shows that the redundancy-resolution problem may be solved in a more favorable manner by using online optimization techniques (e.g., the QP formulation and approach).

In general, there are several extrinsic and inherent factors such as end-effector motion requirement, joint physical limits, and optimization of secondary criteria, which can greatly affect the motion planning of redundant robot manipulators, some allowing kinematic controls to become nonrepetitive [17]. If the redundancy-resolution schemes (e.g., the conventional pseudoinverse-type schemes) are unsuitable for a number of particular end-effector tasks, the final configuration of the robot manipulator may not coincide well with the initial configuration (i.e., the mo-

Y. Zhang, Z. Zhang, *Repetitive Motion Planning and Control of Redundant Robot Manipulators*, DOI 10.1007/978-3-642-37518-7_2,

tion planning of the robot manipulators is not repetitive). The so-called joint-angle drift phenomenon (also referred to as the nonrepetitive problem) implies that, when the robot end-effector tracks a closed path in its workspace, the joint variables may not return to their initial values after completing the end-effector task [11–17]. In other words, the trajectories obtained in the joint space may not be closed. Problems may arise from the robot manipulator's unpredictable behavior and even lead to less efficiency in readjusting the manipulator's configuration with self-motion at every cycle [18, 19]. Thus, it makes good sense if the repetitive motion planning (RMP) of redundant robot manipulators can be achieved by a scheme (e.g., the quadratic-form RMP scheme [12, 13, 15, 17]).

In this chapter, differing from the conventional pseudoinverse-type scheme, an RMP scheme (i.e., joint-angle drift remedy scheme) is presented and investigated for redundant robot manipulators. Note that joint physical limits are incorporated into such an RMP scheme which aims at making the kinematic control repeatable (also referred to as repetitive or cyclic). Then, being the extensions of the first RMP scheme, some other optimization schemes are developed and investigated for the purpose of repetitive motion planning. Finally, these schemes are reformulated and unified as QP problems with different definitions of the same coefficients, which can be solved readily by using recurrent neural networks [1, 3–6, 13, 15, 20] or numerical algorithms based on the linear variational inequality [21–24].

2.2 Physically Constrained RMP Scheme

The pseudoinverse-type solution of the redundancy resolution problem for the redundant robot manipulator, such as (1.3), may not be repetitive [7]. In other words, a closed path of the end-effector may not yield the closed trajectories in the joint space. Such a joint-angle drift is not acceptable for cyclic motion planning and control. The manipulator's configuration can be readjusted with manipulator self-motion processes (i.e., without changing the end-effector's position and orientation) [18, 19], but this would be less efficient. To make the inverse-kinematics solution itself repetitive, the minimization of the joint displacement between current and initial states could be exploited [12, 13, 15, 17]. In the formulation, the performance index used is

$$\frac{1}{2}(\dot{\theta}+\hat{c})^{\mathrm{T}}(\dot{\theta}+\hat{c}) \quad \text{with } \hat{c}=\lambda\big(\theta-\theta(0)\big), \tag{2.1}$$

where $\theta(0)$ is the initial state of the joint-angle vector. The design parameter, $\lambda > 0$, is used to scale the magnitude of the manipulator response to the joint displacement. The design parameter λ should be set as large as the robot system would permit or selected appropriately for simulative and/or experimental purposes. Note that theoretical analysis for the effectiveness of such a performance index (2.1) on RMP of robots will be presented in the next chapter via two different approaches, namely, the gradient-descent method and Zhang et al.'s neural-dynamic method. Now, based on the repetitive performance index (2.1), we have the following basic problem formulation for RMP of redundant robot manipulators:

$$\begin{aligned} &\text{minimize} \quad \frac{1}{2}(\dot{\theta}+\hat{c})^{\mathrm{T}}(\dot{\theta}+\hat{c}) \quad \text{with } \hat{c}=\lambda\big(\theta-\theta(0)\big),\\ &\text{subject to} \quad J(\theta)\dot{\theta}=\dot{r}. \end{aligned}$$

For the above basic RMP scheme, the performance index can be reformulated as $(\dot{\theta}+\hat{c})^{\mathrm{T}}(\dot{\theta}+\hat{c})=\dot{\theta}^{\mathrm{T}}\dot{\theta}+2\hat{c}^{\mathrm{T}}\dot{\theta}+\hat{c}^{\mathrm{T}}\hat{c}$. Since such an RMP scheme is resolved at the joint-velocity level and the decision variable vector is the joint velocity $\dot{\theta}$, the joint angle θ can be regarded as a constant in the minimization of the performance index [i.e., "minimize $(\dot{\theta}+\hat{c})^{\mathrm{T}}(\dot{\theta}+\hat{c})/2$"]. In addition, $\hat{c}^{\mathrm{T}}\hat{c}$ is positive and viewed as a constant (with respect to $\dot{\theta}$), which is thus set aside from "minimize $(\dot{\theta}+\hat{c})^{\mathrm{T}}(\dot{\theta}+\hat{c})/2$". Therefore, the minimization of $(\dot{\theta}+\hat{c})^{\mathrm{T}}(\dot{\theta}+\hat{c})/2$ is equivalent to the minimization of $\dot{\theta}^{\mathrm{T}}\dot{\theta}/2+\hat{c}^{\mathrm{T}}\dot{\theta}$. Besides, in view of the fact that almost all robot manipulators are physically constrained by their joint-angle limits and joint-velocity limits, we have the following more realistic and useful RMP scheme for redundant robot manipulators [12, 13, 15, 17]:

$$\text{minimize} \quad \frac{1}{2}\dot{\theta}^{\mathrm{T}}\dot{\theta}+\hat{c}^{\mathrm{T}}\dot{\theta} \quad \text{with } \hat{c}=\lambda\big(\theta-\theta(0)\big), \tag{2.2}$$

$$\text{subject to} \quad J(\theta)\dot{\theta}=\dot{r}, \tag{2.3}$$

$$\theta^{-}\le\theta\le\theta^{+}, \tag{2.4}$$

$$\dot{\theta}^{-}\le\dot{\theta}\le\dot{\theta}^{+}, \tag{2.5}$$

where (2.3) is exactly (1.2), which describes the end-effector primary task, i.e., the path. More importantly, the avoidance of joint-angle limits $\theta^{\pm}$ and joint-velocity limits $\dot{\theta}^{\pm}$ are considered, with superscripts $^{+}$ and $^{-}$ denoting the upper and lower limits of a joint variable vector (e.g., θ or $\dot{\theta}$), respectively. Since the performance index (2.2) is quadratic, the above optimization scheme (2.2)–(2.5) is hereafter termed quadratic-form physically-constrained RMP scheme for presentation convenience, or termed physically-constrained RMP scheme for simplicity in this book. Lastly, the comparison between the quadratic-form physically constrained RMP scheme and the pseudoinverse-based techniques can be presented as follows [17].

Remark 2.1 Unlike the conventional pseudoinverse-based techniques [2, 7, 8], the physically constrained RMP scheme plans repetitive motion of redundant manipulators in an explicit inverse-free manner, in order to avoid expensive $O(n^3)$ computation. Moreover, physical constraints [i.e., in the form of bound constraints (2.4) and (2.5)] can be incorporated into the presented physically constrained RMP scheme (2.2)–(2.5), whereas pseudoinverse-based techniques do not consider physical constraints; and this is also true for algorithmic singularities [8].

Remark 2.2 Based on the extended Jacobian technique [9, 10], if Jacobian matrix J is a full-rank matrix of dimension $m\times n$, then any matrix P for which $JP=I$

can be obtained by adding appropriately selected $n - m$ constraint-rows to J such that the $n \times n$ extended Jacobian matrix obtained becomes nonsingular. However, such $n - m$ constraints may not take full advantage of redundancy (e.g., they may not permit inequality-based obstacle avoidance [5]) and may artificially limit the end-effector workspace. In addition, the extended Jacobian technique also limits the total number of the physical constraints (i.e., $n - m$), which may have less practical significance in the field of engineering. In contrast, by using the presented physically constrained RMP scheme (2.2)–(2.5) that incorporates the avoidance of joint physical limits [i.e., (2.4) and (2.5)] as a subtask of redundancy-resolution, the proposed framework can also handle other subtasks, such as obstacle avoidance [5], formulated in terms of constraints and/or performance indices.

2.3 Extensions of Physically Constrained RMP Scheme

As mentioned above, the minimization of the joint displacement between current and initial states should be exploited to achieve the RMP purpose. Based on such an idea, some extensions of the above physically constrained RMP scheme (2.2)–(2.5) are discussed, developed, and investigated in this section to remedy the joint-angle drift phenomenon of redundant robot manipulators.

2.3.1 Extension I

In this subsection, the extension-I optimization scheme [being the first extension of the RMP scheme (2.2)–(2.5)] is developed and investigated for RMP of redundant robot manipulators.

The performance index for the extension-I optimization scheme is formulated as

$$\frac{1}{2}\dot{\theta}^{\mathrm{T}} W \dot{\theta} + \hat{c}^{\mathrm{T}} \dot{\theta}, \tag{2.6}$$

where $W := \|\theta - \theta(0)\|_2^2 I$ (with $\|\cdot\|_2$ denoting the two-norm (or termed, Euclidean norm) of a vector, and I denoting the identity matrix), and $\hat{c}$ is defined as before. Note that such a performance index is obtained by defining the scalar-valued error function $\hat{\varepsilon} = \ln(\|\theta - \theta(0)\|_2^2)$ and exploiting the gradient-descent method [16]. Please refer to Appendix A for proof details. Thus, by incorporating the joint physical limits into the scheme formulation, we have the following RMP scheme for physically constrained redundant robot manipulators [16]:

$$\text{minimize} \quad \frac{1}{2}\dot{\theta}^{\mathrm{T}} W \dot{\theta} + \hat{c}^{\mathrm{T}} \dot{\theta} \quad \text{with } \hat{c} = \lambda\big(\theta - \theta(0)\big), \tag{2.7}$$

$$\text{subject to} \quad J(\theta)\dot{\theta} = \dot{r}, \tag{2.8}$$

$$\theta^- \le \theta \le \theta^+, \tag{2.9}$$

$$\dot{\theta}^{-} \leq \dot{\theta} \leq \dot{\theta}^{+}. \tag{2.10}$$

Hereafter, for presentation convenience, the above scheme (2.7)–(2.10) is termed in this book the extension-I optimization scheme (or simply, extension-I scheme) for repetitive motion planning of redundant robot manipulators. The efficacy of the extension-I optimization scheme (2.7)–(2.10) is shown in Appendix B.

2.3.2 Extension II

Facing the success of the aforementioned RMP schemes [i.e., scheme (2.2)–(2.5) and scheme (2.7)–(2.10)] [12, 13, 15–17], we may think of a new performance index for such drift-free purposes, which is formulated as follows [11]:

$$\frac{1}{2}\dot{\theta}^{\mathrm{T}} W \dot{\theta}, \tag{2.11}$$

where W is defined as before, i.e., $W = \|\theta - \theta(0)\|_2^2 I$. Evidently, by minimizing such a performance index (2.11) we can minimize "clearly" the squared norm of joint displacement between current and initial states [in the sense that "when θ equals $\theta(0)$, (2.11) is minimized to be zero; otherwise, it is positive"]. Therefore, with the performance index (2.11) minimized and the joint physical limits considered, the extension-II optimization scheme for the RMP purposes of redundant robot manipulators is formulated as below [11]:

$$\text{minimize} \quad \frac{1}{2}\dot{\theta}^{\mathrm{T}} W \dot{\theta}, \tag{2.12}$$

$$\text{subject to} \quad J(\theta)\dot{\theta} = \dot{r}, \tag{2.13}$$

$$\theta^{-} \leq \theta \leq \theta^{+}, \tag{2.14}$$

$$\dot{\theta}^{-} \leq \dot{\theta} \leq \dot{\theta}^{+}, \tag{2.15}$$

which is the second extension of physically constrained RMP scheme (2.2)–(2.5). It is worth pointing out here that the extension-II optimization scheme (2.12)–(2.15) can be derived from the extension-I optimization scheme (2.7)–(2.10) with design parameter λ being zero. In this sense, the extension-II optimization scheme can be viewed as a special case of the extension-I optimization scheme. However, the inability of extension-II optimization scheme (2.12)–(2.15) is shown in Appendix C.

2.3.3 Extension III

Based on lots of simulative results for the physically constrained RMP scheme [12, 13, 15, 17], we find that the linear part of the performance index (2.2) (i.e., the so-called drift-free criterion $\hat{c}^{\mathrm{T}}\dot{\theta}$) plays an important role in the success of the RMP scheme (2.2)–(2.5) on redundant robot manipulators. Thus, in this subsection, the following linear performance index is exploited in order to achieve the repetitive motion planning of redundant robot manipulators:

$$\hat{c}^{\mathrm{T}}\dot{\theta}, \tag{2.16}$$

where $\hat{c}$ is defined as before, i.e., $\hat{c} = \lambda(\theta - \theta(0))$. Then, similar to the previous subsections, by minimizing the performance index (2.16) and considering the joint physical limits, we have the following redundancy-resolution scheme for the RMP purpose [14]:

$$\text{minimize} \quad \hat{c}^{\mathrm{T}}\dot{\theta} \quad \text{with } \hat{c} = \lambda\big(\theta - \theta(0)\big), \tag{2.17}$$

$$\text{subject to} \quad J(\theta)\dot{\theta} = \dot{r}, \tag{2.18}$$

$$\theta^{-} \leq \theta \leq \theta^{+}, \tag{2.19}$$

$$\dot{\theta}^{-} \leq \dot{\theta} \leq \dot{\theta}^{+}, \tag{2.20}$$

which is called the extension-III optimization scheme in this book for presentation convenience. Note that such a performance index is also the linear part of the performance index (2.7), which has great effect on the success of the extension-I optimization scheme (2.7)–(2.10) (please refer to the simulation part in Appendix B for details) [16]. Evidently, the extension-I optimization scheme can be obtained by combining the extension-II optimization scheme (viewed as the quadratic part) and the extension-III optimization scheme (viewed as the linear part). The interesting results about the extension-III optimization scheme (2.17)–(2.20) are shown in Appendix D.

In summary, being extensions of physically constrained RMP scheme (2.2)–(2.5), these optimization schemes have been developed and discussed in this section for achieving the RMP purpose of redundant robot manipulators.

2.4 QP Reformulation and Unification

In the previous sections, the presented physically constrained RMP scheme (2.2)–(2.5) and its three extensions [i.e., extension-I optimization scheme (2.7)–(2.10), extension-II optimization scheme (2.12)–(2.15), and extension-III optimization scheme (2.17)–(2.20)] have been developed and investigated for RMP of redundant robot manipulators. In this section, these optimization schemes are reformulated and unified as QP problems. Note that the QP unification of these schemes brings more insights into the wealth of existing solutions, as well as a better understanding of future researches (e.g., the unification of various redundancy-resolution schemes via the QP technique).

Since the redundancy-resolution problem is solved at the joint-velocity level, the limited joint-angle range $[\theta^{-}, \theta^{+}]$ (in the presented scheme formulations) has to be

Table 2.1 Definitions of the coefficients Q and $\hat{q}$ for the presented optimization schemes

Scheme	Definition of Q	Definition of $\hat{q}$
Physically constrained RMP scheme (2.2)–(2.5)	$Q := I \in R^{n\times n}$	$\hat{q} := \hat{c} \in R^n$
Extension-I optimization scheme (2.7)–(2.10)	$Q := W \in R^{n\times n}$	$\hat{q} := \hat{c} \in R^n$
Extension-II optimization scheme (2.12)–(2.15)	$Q := W \in R^{n\times n}$	$\hat{q} := 0 \in R^n$
Extension-III optimization scheme (2.17)–(2.20)	$Q := 0 \in R^{n\times n}$	$\hat{q} := \hat{c} \in R^n$

converted into an expression based on the joint velocity $\dot{\theta}$ [12, 13, 15, 17]. The new bound constraint can thus be written as

$$\mu(\theta^- - \theta) \le \dot{\theta} \le \mu(\theta^+ - \theta), \tag{2.21}$$

where large values of parameter $\mu > 0$ may cause quick joint deceleration as the robot approaches its limits [13, 15, 17]. Normally, μ is determined in order for the converted feasible region of $\dot{\theta}$ (i.e., by the joint-angle limits $[\theta^-, \theta^+]$) to be not smaller than the original region made by the joint-velocity limits $[\dot{\theta}^-, \dot{\theta}^+]$. In mathematics, μ should be more than or equal to $2\max_{1\le i\le n}\{\dot{\theta}_i^+/(\theta_i^+ - \theta_i^-), -\dot{\theta}_i^-/(\theta_i^+ - \theta_i^-)\}$ [13, 15, 17].

Combining the bound-constraints (2.21) and $\dot{\theta}^- \le \dot{\theta} \le \dot{\theta}^+$ yields a unified dynamic bound-constraint, $\xi^- \le \dot{\theta} \le \xi^+$, where the ith elements of ξ^- and ξ^+ are defined respectively as [13, 15, 17]:

$$\xi_i^- = \max\{\dot{\theta}_i^-, \mu(\theta_i^- - \theta_i)\}, \qquad \xi_i^+ = \min\{\dot{\theta}_i^+, \mu(\theta_i^+ - \theta_i)\}.$$

Based on the above handling of joint physical limits (i.e., the combination of joint-angle and joint-velocity limits), the presented scheme for RMP purpose can be reformulated and unified as the following quadratic program:

$$\text{minimize} \quad \frac{1}{2}\hat{x}^{\mathrm{T}} Q \hat{x} + \hat{q}^{\mathrm{T}}\hat{x}, \tag{2.22}$$

$$\text{subject to} \quad J\hat{x} = \hat{d}, \tag{2.23}$$

$$\xi^- \le \hat{x} \le \xi^+, \tag{2.24}$$

where the decision variable vector $\hat{x} := \dot{\theta}$ and $\hat{d} = \dot{r}$. Note that the definitions of coefficients $Q \in R^{n\times n}$ and $\hat{q} \in R^n$ for the presented optimization schemes are different from those for each other, which are shown in Table 2.1. So, the quadratic-form or linear-form schemes in the table all can be termed QP-based schemes as they can be reformulated, unified and solved as QPs; and the linear-form extension-III optimization scheme (2.17)–(2.20) can also be termed an LP-based scheme as it can be reformulated and solved as LP (note that LP is a special case of QP and that the linear form is a special case of the quadratic form).

Remark 2.3 As seen from Table 2.1, $Q = I$ and $\hat{q} = \hat{c}$ are set in the QP reformulation for the presented RMP scheme (2.2)–(2.5). Therefore, the coefficient matrix

Q is positive definite, which implies that the objective function in (2.22) is strictly convex. In view of the feasible region made by linear constraints (2.23) and (2.24) being a closed convex set, it follows from [25] that the constrained optimal-solution to QP problem (2.22)–(2.24) [corresponding to the most preferred physically constrained RMP scheme (2.2)–(2.5)] exists uniquely. In light of the uniqueness property, the continuity of the QP solution can thus be guaranteed. It is worth pointing out that the presented unified QP problem (2.22)–(2.24) can be solved by using the recurrent neural networks (e.g., the LVI-based primal–dual neural network) [1, 3–6, 13, 15, 20] and LVI-based numerical algorithms [21–24], which are detailed in Chaps. 4 through 7 of Part II of this book.

2.5 Chapter Conclusions

In this chapter, the physically constrained RMP scheme (2.2)–(2.5) has been presented and investigated to remedy the joint-angle drift phenomena in motion planing of redundant robot manipulators (i.e., to make the kinematic control repeatable). In addition, being the extensions of (2.2)–(2.5), three optimization schemes [i.e., extension-I optimization scheme (2.7)–(2.10), extension-II optimization scheme (2.12)–(2.15), and extension-III optimization scheme (2.17)–(2.20)] have been developed and investigated to achieve the purpose of RMP for redundant robot manipulators. More importantly, these optimization schemes have been finally reformulated and unified as QP problems depicted in (2.22)–(2.24) with different definitions of the same coefficients [i.e., with different Q and $\hat{q}$ in (2.22)].

References

1. Zhang Y (2002) Analysis and design of recurrent neural networks and their applications to control and robotic systems. PhD dissertation, Chinese University of Hong Kong, Hong Kong
2. Siciliano B, Khatib O (2008) Springer handbook of robotics. Springer, Heidelberg
3. Zhang Y, Wang J (2002) A dual neural network for constrained torque optimization of kinematically redundant manipulators. IEEE Trans Syst Man Cybern, Part B, Cybern 32(5):654–662
4. Zhang Y, Wang J, Xia Y (2003) A dual neural network for redundancy resolution of kinematically redundant manipulators subject to joint limits and joint velocity limits. IEEE Trans Neural Netw 14(3):658–667
5. Zhang Y, Wang J (2004) Obstacle avoidance of kinematically redundant manipulators using a dual neural network. IEEE Trans Syst Man Cybern, Part B, Cybern 34(1):752–759
6. Zhang Y, Ge SS, Lee TH (2004) A unified quadratic programming based dynamical system approach to joint torque optimization of physically constrained redundant manipulators. IEEE Trans Syst Man Cybern, Part B, Cybern 34(5):2126–2132
7. Klein CA, Huang C (1983) Review of pseudoinverse control for use with kinematically redundant manipulators. IEEE Trans Syst Man Cybern, Part B, Cybern 13(5):245–250
8. De Luca A, Lanari L, Oriolo G (1992) Control of redundant robots on cyclic trajectories. In: Proceedings of IEEE international conference on robotics and automation, pp 500–506

9. Kireanski MV, Petrovie TM (1993) Combined analytical-pseudoinverse inverse kinematic solution for simple redundant manipulators and singularity avoidance. Int J Robot Res 12(2):188–196
10. Huang L (1997) An extended form of damped pseudoinverse control of kinematically redundant manipulators. In: Proceedings of IEEE international conference on systems, man, and cybernetics, pp 3791–3796
11. Chen K, Guo D, Tan Z, Yang Z, Zhang Y (2008) Cyclic motion planning of redundant robot arms: simple extension of performance index may not work. In: Proceedings of the 2nd international symposium on intelligent information technology application, pp 635–639
12. Zhang Y, Lv X, Li Z, Yang Z, Zhu H (2008) Effective neural remedy for drift phenomenon of planar three-link robot arm using quadratic performance index. Electron Lett 44(6):436–437
13. Zhang Y, Lv X, Li Z, Yang Z, Chen K (2008) Repetitive motion planning of PA10 robot arm subject to joint physical limits and using LVI-based primal-dual neural network. Mechatronics 18(9):475–485
14. Zhang Y, Guo D (2009) Linear programming versus quadratic programming in robots' repetitive redundancy resolution: a chattering phenomenon investigation. In: Proceedings of the 4th IEEE conference on industrial electronics and applications, pp 2822–2827
15. Zhang Y, Tan Z, Chen K, Yang Z, Lv X (2009) Repetitive motion of redundant robots planned by three kinds of recurrent neural networks and illustrated with a four-link planar manipulator's straight-line example. Robot Auton Syst 57(6–7):645–651
16. Zhang Y, Guo D, Cai B, Yi C (2011) Analysis and verification on repetitive motion planning scheme of redundant manipulators using new performance index. J Wuhan Univ Technol 35(1):67–70
17. Zhang Y, Guo D, Cai B, Chen K (2011) Remedy scheme and theoretical analysis of joint-angle drift phenomenon for redundant robot manipulators. Robot Comput-Integr Manuf 27(4):860–869
18. Zhang Y, Zhu H, Tan Z, Cai B, Yang Z (2008) Self-motion planning of redundant robot manipulators based on quadratic program and shown via PA10 example. In: Proceedings of the 2nd international symposium on systems and control in aerospace and astronautics, pp 1–6
19. Zhang Y, Huang Y, Guo D (2009) Self-motion planning of functionally redundant PUMA560 manipulator via quadratic-program formulation and solution. In: Proceedings of IEEE international conference on mechatronics and automation, pp 2518–2523
20. Zhang Y (2005) On the LVI-based primal-dual neural network for solving online linear and quadratic programming problems. In: Proceedings of American control conference, pp 1351–1356
21. He B (1992) A projection and contraction method for a class of linear complementarity problem and its application in convex quadratic programming. Appl Math Optim 25(3):247–262
22. He B (1994) A new method for a class of linear variational inequalities. Math Program 66(1–3):137–144
23. He B (1994) Solving a class of linear projection equations. Numer Math 68(1):71–80
24. Zhang Y, Fu S, Zhang Z, Xiao L, Li X (2011) On the LVI-based numerical method (E47 algorithm) for solving quadratic programming problems. In: Proceedings of IEEE international conference on automation and logistics, pp 125–130
25. Bazarra MS, Sherali HD, Shetty CM (1993) Nonlinear programming—theory and algorithms. Wiley, New York

Chapter 3
Proofs of Repetitive Motion Performance Index

Abstract In this chapter, two kinds of neural network approaches [i.e., gradient neural network (also termed gradient-based neural network), GNN, and Zhang neural network, ZNN] are used to analyze and prove the efficacy of the repetitive motion performance index.

3.1 Proof via GNN Approach

In this section, we employ the well-known gradient-based neural network approach to analyze and prove the repetitive motion performance index.

3.1.1 Background of GNN

The dynamic system approach is one of the important parallel-processing methods for solving online problems. Recently, due to the in-depth research in neural networks, numerous dynamic and analog solvers based on recurrent neural networks (RNNs) have been developed and investigated. The neural dynamic approach is thus regarded as a powerful alternative for online computation because of its parallel distributed nature and convenience of hardware implementation [1]. As one of the important neural dynamic approaches, gradient-based neural network (GNN) is based on the elimination of a scalar-valued norm-based energy function [2]. In this chapter, we first analyze and prove repetitive motion performance index (2.1) based on the gradient-based neural network (GNN) approach [1, 2]. For the convenience of discussion, the RMP performance index can be repeated below:

$$\frac{1}{2}(\dot{\theta}+\hat{c})^{\mathrm{T}}(\dot{\theta}+\hat{c}) \quad \text{with } \hat{c}=\lambda\big(\theta-\theta(0)\big), \tag{3.1}$$

where $\theta(0)$ is the initial state of the joint variable vector. The design parameter, $\lambda > 0$, is used to scale the magnitude of the manipulator response to the joint displacement.

Y. Zhang, Z. Zhang, *Repetitive Motion Planning and Control of Redundant Robot Manipulators*, DOI 10.1007/978-3-642-37518-7_3,

3.1.2 Proof

To derive repetitive motion performance index (3.1), we can take the following steps via the so-called GNN approach.

First, to achieve a repetitive motion of both end-effector and joint angles [i.e., $\theta(T_{\text{final}}) = \theta(0)$, where T_{final} denotes the final time of the repetitive motion], it is natural to define the following joint-change function of scalar-valued two-norm-based form: $\varepsilon = \|\theta(t) - \theta(0)\|_2^2/2$, which is to be minimized over the cyclic task duration $[0, T_{\text{final}}]$. Evidently, the repetitive purpose is achieved if the minimum zero value of this joint-change function ε is finally obtained at time instant $t = T_{\text{final}}$ [i.e., if $\theta(T_{\text{final}}) - \theta(0) = 0$].

Second, using the GNN approach [1, 3], we set $\dot{\theta} = -\lambda(\partial\varepsilon/\partial\theta)$, i.e., $\dot{\theta} = -\lambda(\theta - \theta(0))$. In addition, the above can be rewritten as $\dot{\theta} + \lambda(\theta - \theta(0)) = 0$.

Finally, as other factors (e.g., end-effector motion requirement and avoidance of joint physical limits) have to be considered in the RMP [4, 5], the dynamic equation $\dot{\theta}(t) + \lambda(\theta(t) - \theta(0)) = 0$ could be achieved only theoretically. Thus, minimizing $\|\dot{\theta}(t) + \lambda(\theta(t) - \theta(0))\|_2^2/2$ appears to be more feasible in practice for the repetitive motion of redundant manipulators. It follows that expanding $\|\dot{\theta}(t) + \lambda(\theta(t) - \theta(0))\|_2^2/2$ yields

$$\big(\dot{\theta} + \lambda\big(\theta(t) - \theta(0)\big)\big)^{\mathrm{T}}\big(\dot{\theta} + \lambda\big(\theta(t) - \theta(0)\big)\big)/2.$$

With $\hat{c} := \lambda(\theta(t) - \theta(0))$, therefore, the above can be written as $(\dot{\theta} + \hat{c})^{\mathrm{T}}(\dot{\theta} + \hat{c})/2$, which is exactly the same as repetitive motion performance index (3.1).

In summary, by using the GNN approach, the proof and the theoretical effectiveness of drift-free performance index $(\dot{\theta} + \hat{c})^{\mathrm{T}}(\dot{\theta} + \hat{c})/2$ used in the repetitive motion planning of redundant robot manipulators have been demonstrated.

3.2 Proof via ZNN Approach

Different from conventional gradient neural networks (GNN), the design of Zhang neural network (ZNN) is based on a vector/matrix-valued error function, instead of the scalar-valued norm-based energy function. For comparison, for illustration, and for completeness, in this section, we employ an alternative approach [i.e., Zhang neural network (ZNN)] to prove the repetitive motion performance index (3.1) theoretically.

3.2.1 Background of ZNN

Since March 2001 [6, 7], Zhang et al. have formally proposed a special class of recurrent neural networks, termed Zhang neural network (ZNN) for online time-varying problems solving (e.g., time-varying linear equation solving and time-varying matrix inversion) [7]. Note that the design procedure of the ZNN method can be further used for proving the repetitive motion performance index for redundant robot manipulators [5, 8]. In this section, we prove the repetitive motion performance index (3.1) via ZNN approach.

3.2.2 Proof

The proof via this approach is shown as follows.

First, to eliminate the joint-angle drift phenomenon, we define alternatively the following vector-valued displacement-function, which is to be eliminated over the whole range $[0, T_{\text{final}}]$:

$$E(t) = \theta(t) - \theta(0) \in R^n.$$

Note that $E(t) = 0$ if and only if (iff, for short) $\theta - \theta(0) = 0$.

Second, in order for $E(t)$ to converge exponentially to 0, we simply set

$$\dot{E}(t) = -\lambda E(t), \tag{3.2}$$

where $\lambda > 0 \in R$ denotes the exponential convergence rate of E. The solution to Eq. (3.2) is evidently $E(t) = \exp(-\lambda t)E(0)$. This also suggests that within the time-period of $4/\lambda$ seconds, $|E_j(t)|$ would be less than 1.85 % of $|E_j(0)|$ for all $j \in \{1, 2, \ldots, n\}$ [6].

Finally, substituting $E(t) = \theta(t) - \theta(0)$ into (3.2) yields $\dot{\theta}(t) = -\lambda(\theta(t) - \theta(0))$. That is, $\dot{\theta}(t) + \lambda(\theta(t) - \theta(0)) = 0$. Inasmuch as the other factors like the end-effector motion requirement and joint physical limits have to be considered in the RMP [5, 8], the equation $\dot{\theta}(t) + \lambda(\theta(t) - \theta(0)) = 0$ can only be achieved theoretically. Thus, minimizing $\|\dot{\theta}(t) + \lambda(\theta(t) - \theta(0))\|_2^2/2$ appears to be more feasible for the RMP of redundant manipulators, rather than forcing $\dot{\theta}(t) + \lambda(\theta(t) - \theta(0)) = 0$ directly in its exact form. Expanding $\|\dot{\theta}(t) + \lambda(\theta(t) - \theta(0))\|_2^2/2$ yields

$$\big(\dot{\theta} + \lambda\big(\theta(t) - \theta(0)\big)\big)^{\mathrm{T}}\big(\dot{\theta} + \lambda\big(\theta(t) - \theta(0)\big)\big)/2. \tag{3.3}$$

Thereafter, with $\hat{c} := \lambda(\theta(t) - \theta(0))$, the performance index (3.3) becomes exactly (3.1); that is, $(\dot{\theta} + \hat{c})^{\mathrm{T}}(\dot{\theta} + \hat{c})/2$, which has been proven effective for the drift-free motion planning of redundant robots.

Remark 3.1 The ZNN and GNN approaches are two different kinds of neural-network methods though the results being the same in our RMP context. The comparisons between the ZNN and GNN approaches are presented as follows. First, the design of GNN is based on the elimination of a scalar-valued two-norm joint-change function. In contrast, the design of ZNN is based on the elimination of every entry of a vector-valued displacement-function. Second, as a new model of recurrent neural network, the ZNN model is depicted generally in an implicit dynamics, whereas the GNN model is depicted generally in an explicit dynamics. Finally, the ZNN approach could be guaranteed to globally exponentially converge to the theoretical solution of the time-varying QP problem. In contrast, the GNN approach could only approximately approach the time-varying solution, even when a large enough design parameter is used. Evidently, the exact convergent properties of the two models are different.

3.3 Chapter Conclusions

In this chapter, via the GNN and ZNN approaches, we have shown the design method, theoretical correctness and effectiveness of performance index $(\dot{\theta}+\hat{c})^{\mathrm{T}}(\dot{\theta}+\hat{c})/2$ employed in repetitive motion planning of redundant robot manipulators.

References

1. Zhang Y, Chen K (2008) Global exponential convergence and stability of Wang neural network for solving online linear equations. Electron Lett 44(2):145–146
2. Zhang Y, Chen K, Tan H (2009) Performance analysis of gradient neural network exploited for online time-varying matrix inversion. IEEE Trans Autom Control 54(8):1940–1945
3. Zhang Y, Lv X, Li Z, Yang Z, Zhu H (2008) Effective neural remedy for drift phenomenon of planar three-link robot arm using quadratic performance index. Electron Lett 44(6):435–437
4. Zhang Y, Guo D, Cai B, Chen K (2011) Remedy scheme and theoretical analysis of joint-angle drift phenomenon for redundant robot manipulators. Robot Comput-Integr Manuf 27(4):860–869
5. Chen K, Zhang L, Zhang Y (2008) Cyclic motion generation of multi-link planar robot performing square end-effector trajectory analyzed via gradient-descent and Zhang et al.'s neural-dynamic methods. In: Proceedings of international symposium on systems and control in aerospace and astronautics, pp 1–6
6. Zhang Y (2002) Analysis and design of recurrent neural networks and their applications to control and robotic systems. PhD dissertation, Chinese University of Hong Kong, Hong Kong
7. Zhang Y, Jiang D, Wang J (2002) A recurrent neural network for solving Sylvester equation with time-varying coefficients. IEEE Trans Neural Netw 13(5):1053–1063
8. Zhang Y, Guo D, Cai B, Chen K (2011) Remedy scheme and theoretical analysis of joint-angle drift phenomenon for redundant robot manipulators. Robot Comput-Integr Manuf 27(4):860–869

Part II
QP Solvers

Chapter 4
Dual Neural Network

Abstract One of state-of-the-art recurrent neural networks (RNNs) is dual neural network (DNN). It can solve QP in real time. The dual neural network is of simple piecewise-linear dynamics and has global (exponential) convergence to optimal solutions. In this chapter, we first introduce the QP problem formulation and its online solution based on RNN. Some related concepts and definitions are also given. Second, we present the DNN and its design method. In addition to the general design method, for nondiagonal, nonanalytical, and/or time-varying cases, a matrix-inverse neural network could be combined into such a design procedure of the DNN for online computation of its matrix-inverse related term. Third, we show the analysis results of the DNN. Finally, we present a numerical simulation and illustrative example of using the DNN to solve static QP problems.

4.1 Introduction

In the 1980s, Hopfield and Tank published their working neural networks implemented on analogue circuits and designed for solving optimization problems [1]. Since then, the RNN approach has been thought to be a powerful alternative to real-time optimization. This is in light of the parallel computing nature and hardware-implementation feasibility of the dynamic-system solvers.

In the past two decades, various RNN models have been developed for solving the quadratic-programming problems. For example, those based on the penalty-parameter method [2], the Lagrangian method [3, 4], the traditional primal–dual method [5, 6], the dual method [7–9], and the LVI (linear variational inequality) based primal–dual method [10–14]. The following facts are known.

- The penalty-based neural network model [2] contains finite penalty parameters and thus generates approximate solutions only.
- When solving inequality-constrained QPs, the Lagrangian neural network [3] may exhibit the premature defect [4]. In addition, the dimensionality of the Lagrangian neural network is much larger than that of original (also termed primal) problems, due to the introduction of slack and surplus variables.
- As a much flexible tool for exactly solving QPs, the traditional primal–dual neural networks [5, 6] were developed with the feature that they handle the origi-

Y. Zhang, Z. Zhang, *Repetitive Motion Planning and Control of Redundant Robot Manipulators*, DOI 10.1007/978-3-642-37518-7_4,

nal/primal quadratic program and its dual problem simultaneously by minimizing the duality gap with KKT conditions and gradient method. Unfortunately, the dynamic equations of the traditional primal–dual neural network are usually complex and may contain high-order nonlinear terms.

- The dual neural network [7–9] is one of state-of-the-art RNNs, designed based on the reformulation of a hybrid-constraint QP problem to a fully bound-constrained QP problem. It has simple piecewise linear dynamics and global (exponential) convergence to optimal solutions of QP problems.
- The LVI-based primal–dual neural network [10, 13] is a state-of-the-art neural model as well, designed based on the reformulation of QP/LP problems to the LVI problems. It has simple piecewise linear dynamics and global (exponential) convergence to optimal solutions of QP and LP problems.

4.1.1 Concepts and Definitions

For a better understanding of the naming of recurrent neural networks, the following concepts are presented that have been scattered in the literature.

Concept 4.1 A recurrent neural network is called a primal neural network if the network dynamic equation and implementation only use the original (also termed primal) decision variables (here, the original/primal decision vector is $\hat{x}$). For example, some neural models in [1, 2, 7, 15] are primal neural networks.

Concept 4.2 A recurrent neural network is called a nonprimal neural network if the network dynamic equation and implementation use a group of auxiliary decision variables, in addition to using the original/primal ones or not. For example, nonprimal neural networks could be the primal–dual neural networks in [5, 6, 10–14], the dual neural networks in [7–9], or the Lagrangian neural networks in [3, 4].

Concept 4.3 A recurrent neural network is called a primal–dual neural network if the network dynamic equation and implementation use a group of dual decision variables, in addition to using the original/primal ones. For example, some neural models in [5, 6, 10–14] are primal–dual neural networks.

Concept 4.4 A recurrent neural network is called a dual neural network if the network dynamic equation and implementation only use a group of dual decision variables. For example, some neural models in [7–9] are dual neural networks.

For a better understanding of the solution characteristics of recurrent neural networks, the following definitions are presented that could have been scattered in the literature (see [7, 8, 13, 16–18] and the references therein).

Definition 4.1 A recurrent neural network is said to be globally convergent if starting from any initial point $\hat{x}(t_0)$ taken in the whole associated Euclidean space, every state trajectory of the recurrent neural network converges to an equilibrium point $\hat{x}^*$ that depends on the initial state of the trajectory. Note that the initial time instant $t_0 \geq 0$, and there might exist many nonisolated equilibrium points $\hat{x}^*$.

Definition 4.2 A recurrent neural network is said to be globally exponentially convergent if every trajectory starting from any initial point $\hat{x}(t_0)$ satisfies: $\forall t \geq t_0 \geq 0$,

$$\left\|\hat{x}(t) - \hat{x}^*\right\| \leq \eta \left\|\hat{x}(t_0) - \hat{x}^*\right\| \exp\bigl(-\lambda(t - t_0)\bigr),$$

where η and λ are positive constants, $\hat{x}^*$ is an equilibrium point depending on initial states $\hat{x}(t_0)$, and the symbol $\|\cdot\|$ hereafter denotes the two-norm of a vector or the Frobenius norm of a matrix unless specified otherwise. Note that there might exist many nonisolated equilibrium points $\hat{x}^*$, and the exponential convergence implies the arbitrarily fast convergence of the recurrent neural network.

Definition 4.3 An equilibrium point $\hat{x}^*$ of a recurrent neural network is locally stable if every trajectory, starting from the initial condition $\hat{x}(t_0)$ near the equilibrium point $\hat{x}^*$, stays near that equilibrium point $\hat{x}^*$. On the other hand, an equilibrium point $\hat{x}^*$ of a recurrent neural network is globally stable if every trajectory starting from any initial condition $\hat{x}(t_0)$ stays near that equilibrium point $\hat{x}^*$.

Definition 4.4 An equilibrium point $\hat{x}^*$ of a recurrent neural network is locally asymptotically stable if it is locally stable and, in addition, the state $\hat{x}(t)$ of the recurrent neural network, starting from any initial condition $\hat{x}(t_0)$ near the equilibrium point $\hat{x}^*$, converges to the equilibrium point $\hat{x}^*$ as time t increases. Note that the equilibrium point $\hat{x}^*$ is assumed here to be isolated.

Definition 4.5 An equilibrium point $\hat{x}^*$ of a recurrent neural network is globally asymptotically stable if it is globally stable and, in addition, the state $\hat{x}(t)$ of the recurrent neural network converges to the equilibrium point $\hat{x}^*$ as time t increases. Note that the equilibrium point $\hat{x}^*$ is assumed here to be unique.

Definition 4.6 An equilibrium point $\hat{x}^*$ of a recurrent neural network is locally exponentially stable if it is locally asymptotically stable and, in addition, the state $\hat{x}(t)$ of the recurrent neural network, starting from any initial condition $\hat{x}(t_0)$ near the equilibrium point $\hat{x}^*$, converges as follows to the equilibrium point $\hat{x}^*$ as time t increases: $\forall t \geq t_0 \geq 0$,

$$\left\|\hat{x}(t) - \hat{x}^*\right\| \leq \eta \left\|\hat{x}(t_0) - \hat{x}^*\right\| \exp\bigl(-\lambda(t - t_0)\bigr),$$

where η and λ are positive constants, and $\hat{x}^*$ is an isolated equilibrium point.

Definition 4.7 An equilibrium point $\hat{x}^*$ of a recurrent neural network is globally exponentially stable if it is globally asymptotically stable and, in addition, the state

$\hat{x}(t)$ of the recurrent neural network converges as follows to the equilibrium point $\hat{x}^*$ as time t increases: $\forall t \geq t_0 \geq 0$,

$$\left\|\hat{x}(t) - \hat{x}^*\right\| \leq \eta \left\|\hat{x}(t_0) - \hat{x}^*\right\| \exp\bigl(-\lambda(t - t_0)\bigr),$$

where η and λ are positive constants, and $\hat{x}^*$ is a unique equilibrium point.

Remark 4.1 Note that, in the context of recurrent neural networks, the global (exponential) convergence of a neural system does not imply the global (exponential) stability of such a neural system. This is because usually there might exist an infinite number of nonisolated solutions to an optimization problem (e.g., in the LP problem or in the QP problem with $\hat{Q}$ positive semidefinite, i.e., $\hat{Q} \succeq 0$) [19]. If all the state trajectories of a recurrent neural network converge to one of such (nonisolated) solutions/equilibriums, we call the recurrent neural network globally convergent. On the other hand, the conventional global stability of a system requires the equilibrium point to be unique (and also isolated).

Remark 4.2 Strictly speaking, we should seldom say that a recurrent neural network (being a dynamic system) is stable or unstable. Instead, we talk about an equilibrium point (of such a neural network system) being stable or not. Thus, we first find the equilibrium points of such a neural network system and then determine whether those equilibrium points are stable or not. In the case that a neural network system has a single equilibrium point that is globally attractive (meaning that all solutions converge to that equilibrium point from all initial conditions), we sometimes get sloppy, saying that the recurrent neural network system is (globally) stable. But we really should say that the equilibrium point is globally stable. On the other hand, this comment is quite different from the usage of "global (exponential) convergence of a recurrent neural network," which is discussed in Definitions 4.1 and 4.2, and Remark 4.1. Moreover, the concept/definition of global (exponential) stability may be extended in the future if necessary.

4.1.2 Primal and Dual Problems

In the design of the traditional primal–dual neural networks [5, 6], the dual neural networks [7–9], and the LVI-based primal–dual neural networks [10–14], when we solve the original/primal QP problem (4.2)–(4.5), we are implicitly solving its dual QP problem (or at least use its dual decision variables). So, it is worth spending a little time and text here on the introduction of the dual problem, dual decision vectors, and duality concept in the context of optimization.

For the LP situation as an example [see the LP converted from (2.17)–(2.20)], the primal and dual problems could be easily written down [19]. For example, if the primal LP problem is

$$\begin{aligned}
&\text{minimize } \hat{q}^{\mathrm{T}}\hat{x},\\
&\text{subject to } J\hat{x} = \hat{d},\\
&\qquad\qquad A\hat{x} \le \hat{b},\\
&\qquad\qquad \xi^- \le \hat{x} \le \xi^+,
\end{aligned}$$

then its dual LP problem is

$$\begin{aligned}
&\text{maximize } \hat{d}^{\mathrm{T}}u - \hat{b}^{\mathrm{T}}v + \xi^{-\mathrm{T}}\underline{\nu} - \xi^{+\mathrm{T}}\overline{\nu},\\
&\text{subject to } J^{\mathrm{T}}u - A^{\mathrm{T}}v + \underline{\nu} - \overline{\nu} = \hat{q},\\
&\qquad u \text{ unrestricted}, \quad v \ge 0,\ \underline{\nu} \ge 0,\ \overline{\nu} \ge 0,
\end{aligned}$$

where

- dual decision vector $u \in R^m$ corresponds to the equality constraint $J\hat{x} = \hat{d}$;
- dual decision vector $v \in R^{\dim(\hat{b})}$ corresponds to the inequality constraint $A\hat{x} \le \hat{b}$, with $\dim(\cdot)$ denoting the dimension of a vector;
- dual decision vector $\underline{\nu} \in R^n$ corresponds to the left part of the bound constraint $\xi^- \le \hat{x} \le \xi^+$, i.e., $\xi^- \le \hat{x}$; and
- dual decision vector $\overline{\nu} \in R^n$ corresponds to the right part of the bound constraint $\xi^- \le \hat{x} \le \xi^+$, i.e., $\hat{x} \le \xi^+$.

Remark 4.3 A minimizing problem can be reformulated as a maximizing problem by changing the sign of the objective function; e.g., "to maximize $\hat{d}^{\mathrm{T}}u - \hat{b}^{\mathrm{T}}v + \xi^{-\mathrm{T}}\underline{\nu} - \xi^{+\mathrm{T}}\overline{\nu}$" equals "to minimize $-\hat{d}^{\mathrm{T}}u + \hat{b}^{\mathrm{T}}v - \xi^{-\mathrm{T}}\underline{\nu} + \xi^{+\mathrm{T}}\overline{\nu}$." Similarly, any $\ge$ inequality can be reformulated as a $\le$ inequality by multiplying -1 on both sides; e.g., $A\hat{x} \le \hat{b}$ equals $-A\hat{x} \ge -\hat{b}$. It is a little bit particular to deal with equality constraints, e.g., $J\hat{x} = \hat{d}$. If the ith constraint in the primal problem is an equation, then the ith variable in the dual problem should be unrestricted. That is why we have the statement "dual decision vector $u \in R^m$ corresponds to the equality constraint $J\hat{x} = \hat{d}$, and u is unrestricted."

For the quadratic-programming situation as another example, we first show a simple example on how to derive the dual problem from its primal problem [19, 20] and then give a theorem on the general QP situation. For instance, if the primal QP problem is

$$\begin{aligned}
&\text{minimize } \frac{1}{2}\hat{x}^{\mathrm{T}}\hat{Q}\hat{x} + \hat{q}^{\mathrm{T}}\hat{x},\\
&\text{subject to } A\hat{x} \le \hat{b},
\end{aligned}$$

we have its Lagrangian dual problem as follows. That is, to maximize $\inf_{\hat{x}\in R^n} L(\hat{x}, v)$ over $v \ge 0$, where dual decision vector $v \in R^{\dim(\hat{b})}$ corresponds to the inequality constraint $A\hat{x} \le \hat{b}$, and

$$L(\hat{x}, v) = \frac{1}{2}\hat{x}^{\mathrm{T}}\hat{Q}\hat{x} + \hat{q}^{\mathrm{T}}\hat{x} + v^{\mathrm{T}}(A\hat{x} - \hat{b}).$$

Note that, for a given v, the function $L(\hat{x}, v)$ is convex, and thus a necessary and sufficient condition for a minimum is that the gradient vanishes to zero, i.e., $\hat{Q}\hat{x} + A^{\mathrm{T}}v + \hat{q} = 0$. Hence, the dual QP problem can be written as follows:

$$\begin{aligned} &\text{maximize } \frac{1}{2}\hat{x}^{\mathrm{T}}\hat{Q}\hat{x} + \hat{q}^{\mathrm{T}}\hat{x} + v^{\mathrm{T}}(A\hat{x} - \hat{b}), \\ &\text{subject to } \hat{Q}\hat{x} + A^{\mathrm{T}}v + \hat{q} = 0, \quad v \ge 0. \end{aligned} \tag{4.1}$$

Now, starting from $\hat{Q}\hat{x} + A^{\mathrm{T}}v + \hat{q} = 0$, we have $-\hat{x}^{\mathrm{T}}\hat{Q}\hat{x} = \hat{x}^{\mathrm{T}}A^{\mathrm{T}}v + \hat{x}^{\mathrm{T}}\hat{q}$ by pre-multiplying $-\hat{x}^{\mathrm{T}}$. Substituting this new equation into (4.1), we can derive the following dual quadratic-programming problem:

$$\begin{aligned} &\text{maximize } -\frac{1}{2}\hat{x}^{\mathrm{T}}\hat{Q}\hat{x} - v^{\mathrm{T}}\hat{b}, \\ &\text{subject to } \hat{Q}\hat{x} + A^{\mathrm{T}}v + \hat{q} = 0, \quad v \ge 0. \end{aligned}$$

In the same manner, we can derive the following theorem.

Theorem 4.1 *For the primal QP problem*

$$\textit{minimize } \hat{x}^{\mathrm{T}}\hat{Q}\hat{x}/2 + \hat{q}^{\mathrm{T}}\hat{x}, \tag{4.2}$$

$$\textit{subject to } J\hat{x} = \hat{d}, \tag{4.3}$$

$$A\hat{x} \le \hat{b}, \tag{4.4}$$

$$\xi^{-} \le \hat{x} \le \xi^{+}, \tag{4.5}$$

its dual QP problem can be derived as

$$\begin{aligned} &\textit{maximize } -\frac{1}{2}\hat{x}^{\mathrm{T}}\hat{Q}\hat{x} + \hat{d}^{\mathrm{T}}u - \hat{b}^{\mathrm{T}}v + \xi^{-\mathrm{T}}\underline{v} - \xi^{+\mathrm{T}}\overline{v}, \\ &\textit{subject to } \hat{Q}\hat{x} + \hat{q} - J^{\mathrm{T}}u + A^{\mathrm{T}}v - \underline{v} + \overline{v} = 0, \\ &\qquad u \textit{ unrestricted}, \quad v \ge 0,\ \underline{v} \ge 0,\ \overline{v} \ge 0, \end{aligned}$$

where the dual decision vector $u \in R^m$ corresponds to the equality constraint $J\hat{x} = \hat{d}$; the dual decision vector $v \in R^{\dim(\hat{b})}$ corresponds to the inequality constraint $A\hat{x} \le \hat{b}$; the dual decision vector $\underline{v} \in R^n$ corresponds to the left part of the bound constraint $\xi^{-} \le \hat{x} \le \xi^{+}$, i.e., $\xi^{-} \le \hat{x}$; and the dual decision vector $\overline{v} \in R^n$ corresponds to the right part of the bound constraint $\xi^{-} \le \hat{x} \le \xi^{+}$, i.e., $\hat{x} \le \xi^{+}$.

Proof It can be obtained from [13, 19] or by following the previous derivation. □

Remark 4.4 Note that, to solve (4.2)–(4.5), the usual neural network design methods convert the equality constraint and two-sided inequality constraint into two one-sided inequality constraints. This will unnecessarily increase the dimension of the dual space and thus introduce some excessive neurons. In comparison, if we can

- treat the equality constraint $J\hat{x} = \hat{d}$ in (4.3) as a special two-sided bound constraint (with the lower bound equal to the upper bound, i.e., $\hat{d} = \hat{d}$) and
- treat the one-sided inequality constraint $A\hat{x} \leq \hat{b}$ in (4.4) as a two-sided bound constraint (with the lower bound equal to "the negative infinity" $-\infty$).

This may provide a unified treatment of both equality, inequality, and simple bound constraints. The resultant dual-neural-network (DNN) model for solving general-form quadratic program (4.2)–(4.5) will be of size $n + m + \dim(\hat{b})$ only.

4.2 Dual-Neural-Network Design

To solve online the QP problem depicted in (4.2)–(4.5) via a dual neural network, the following design procedure is presented [7, 9].

4.2.1 The First Step

First, as discussed in Remark 4.4, we treat the equality constraint $J\hat{x} = \hat{d}$ and inequality constraint $A\hat{x} \leq \hat{b}$ respectively in (4.3) and (4.4) as special cases of the augmented bound constraint $\zeta^- \leq H\hat{x} \leq \zeta^+$. To do so, the lower bound ζ^- and upper bound ζ^+ and augmentation matrix H are defined as follows:

$$\zeta^- := \begin{pmatrix} \hat{d} \\ -\varpi \\ \xi^- \end{pmatrix} \in R^{m+\dim(\hat{b})+n}, \qquad \zeta^+ := \begin{pmatrix} \hat{d} \\ \hat{b} \\ \xi^+ \end{pmatrix},$$

$$H := \begin{pmatrix} J \\ A \\ I \end{pmatrix} \in R^{(m+\dim(\hat{b})+n)\times n},$$

where $\varpi \gg 0 \in R^{\dim(\hat{b})}$ is defined sufficiently large to replace $+\infty \in R^{\dim(\hat{b})}$ numerically for simulation and hardware-implementation purposes. In this sense, the QP problem in (4.2)–(4.5) is converted to the following bounded QP problem:

$$\text{minimize } \hat{x}^{\mathrm{T}}\hat{Q}\hat{x}/2 + \hat{q}^{\mathrm{T}}\hat{x}, \tag{4.6}$$

$$\text{subject to } \zeta^- \leq H\hat{x} \leq \zeta^+. \tag{4.7}$$

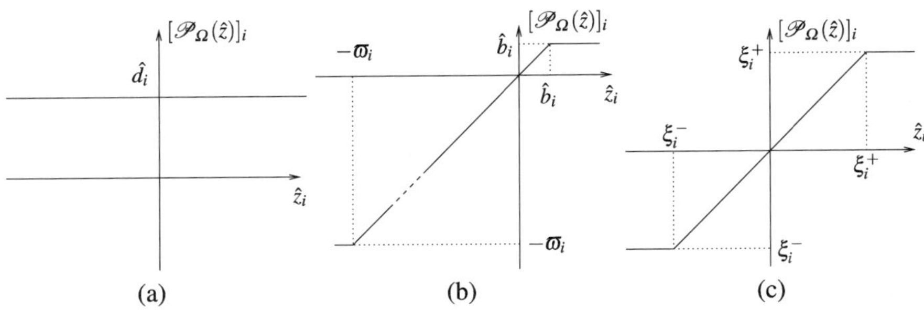

Fig. 4.1 The ith scalar-input scalar-output processing element of the vector-input vector-output projection operator $\mathscr{P}_\Omega(\hat{z})$ from $\hat{z}_i$ onto $[\zeta_i^-, \zeta_i^+]$: **(a)** $i = 1, 2, \ldots, m$, **(b)** $i = (m+1), (m+2), \ldots, (m + \dim(\hat{b}))$, **(c)** $i = (m + \dim(\hat{b}) + 1), (m + \dim(\hat{b}) + 2), \ldots, (m + \dim(\hat{b}) + n)$

4.2.2 The Second Step

The second step of the dual-neural-network design is that we can then reformulate the KKT optimality/complementarity conditions of bound-constraint QP problem (4.6)–(4.7) to a system of piecewise-linear equations [7–9].

Before presenting the following theorem, we define

- a (final) dual decision vector $\hat{z} \in R^{m+\dim(\hat{b})+n}$ corresponding to the augmented bound constraint (4.7), i.e., $\zeta^- \leq H\hat{x} \leq \zeta^+$; and
- a vector-input vector-output projection function/operator $\mathscr{P}_\Omega(\hat{z}) = [[\mathscr{P}_\Omega(\hat{z})]_1, [\mathscr{P}_\Omega(\hat{z})]_2, \ldots, [\mathscr{P}_\Omega(\hat{z})]_{m+\dim(\hat{b})+n}]^{\mathrm{T}}$ from $\hat{z} \in R^{m+\dim(\hat{b})+n}$ onto projection set $\Omega := [\zeta^-, \zeta^+]$, with its processing element $[\mathscr{P}_\Omega(\hat{z})]_i$ being defined hereafter as

$$[\mathscr{P}_\Omega(\hat{z})]_i = \begin{cases} \zeta_i^- & \text{if } \hat{z}_i < \zeta_i^-, \\ \hat{z}_i & \text{if } \zeta_i^- \leq \hat{z}_i \leq \zeta_i^+, \\ \zeta_i^+ & \text{if } \hat{z}_i > \zeta_i^+. \end{cases} \quad i = 1, 2, \ldots, \big(m + \dim(\hat{b}) + n\big),$$

Figure 4.1 shows three situations of the ith scalar-input scalar-output processing element of projection operator $\mathscr{P}_\Omega(\hat{z})$ from $\hat{z}_i$ onto the projection set $[\zeta_i^-, \zeta_i^+]$. That is, graph (a) of Fig. 4.1 is for the equality constraint $J\hat{x} = \hat{d}$, graph (b) is for the one-sided inequality constraint $A\hat{x} \leq \hat{b}$, and graph (c) is for simple bound constraint $\xi^- \leq \hat{x} \leq \xi^+$.

Theorem 4.2 *$\hat{x}$ is an optimal solution to the bound-constraint QP problem depicted in* (4.6) *and* (4.7) *if and only if there exists the (final) dual decision vector $\hat{z} \in R^{m+\dim(\hat{b})+n}$ such that $\hat{Q}\hat{x} - H^{\mathrm{T}}\hat{z} + \hat{q} = 0$ and $H\hat{x} = \mathscr{P}_\Omega(H\hat{x} - \hat{z})$.*

Proof It follows from the KKT conditions [7, 9, 13, 19] that $\hat{x}$ is an optimal solution to the bound-constraint QP problem in (4.6) and (4.7) if and only if there exists a dual decision vector $\hat{z} \in R^{m+\dim(\hat{b})+n}$ such that $\hat{Q}\hat{x} - H^{\mathrm{T}}\hat{z} + \hat{q} = 0$ and

$$\begin{cases} [H\hat{x}]_i = \zeta_i^- & \text{if } \hat{z}_i > 0, \\ [H\hat{x}]_i = \zeta_i^+ & \text{if } \hat{z}_i < 0, \\ \zeta_i^- \le [H\hat{x}]_i \le \zeta_i^+ & \text{if } \hat{z}_i = 0. \end{cases} \tag{4.8}$$

Equation (4.8) is termed the complementarity condition, which is equivalent to the system of piecewise linear equations, $H\hat{x} = \mathscr{P}_\Omega(H\hat{x} - \hat{z})$ [21–23], where $\mathscr{P}_\Omega(\hat{z})$ has been defined before. Therefore, $\hat{x}$ is an optimal solution to the bound-constraint QP problem depicted in (4.6) and (4.7) if and only if there exists the final dual decision vector $\hat{z} \in R^{m+\dim(\hat{b})+n}$ such that $\hat{Q}\hat{x} - H^{\mathrm{T}}\hat{z} + \hat{q} = 0$ and $H\hat{x} = \mathscr{P}_\Omega(H\hat{x} - \hat{z})$. For a better understanding, please also refer to Sect. 4.1.2 about the primal and dual QP problems. □

Considering that coefficient matrix $\hat{Q}$ in (4.2) is assumed to be positive definite in this chapter, it follows immediately from Theorem 4.2 that $\hat{x}$ is an optimal solution to the bound-constraint QP problem depicted in (4.6) and (4.7) [which is exactly the same as the QP problem formulation depicted in (4.2)–(4.5)] if and only if there exists a dual decision vector $\hat{z} \in R^{m+\dim(\hat{b})+n}$ such that

$$\begin{cases} \mathscr{P}_\Omega(HMH^{\mathrm{T}}\hat{z} - HM\hat{q} - \hat{z}) - (HMH^{\mathrm{T}}\hat{z} - HM\hat{q}) = 0, \\ \hat{x} = MH^{\mathrm{T}}\hat{z} - M\hat{q}, \end{cases} \tag{4.9}$$

where, for notational simplicity, we use $M \in R^{n\times n}$ to denote the inverse $\hat{Q}^{-1}$. Here, we may assume that $\hat{Q}$ (and M) is diagonal or we have the analytical form of M for faster computation. The following three examples are given for a better understanding of such an assumption.

- In the minimum-velocity-norm and minimum-acceleration-norm motion planning [7, 10], the coefficient matrix $\hat{Q}$ is the identity matrix I, and so is M.
- In the bi-criteria control of minimum-effort and minimum-energy motion planning [7, 17], the coefficient matrix $\hat{Q}$ is a diagonal matrix of the form of $\mathrm{diag}(\tilde{\alpha}I, 1-\tilde{\alpha})$ with $\tilde{\alpha} \in (0,1)$ here being a weighting factor, and M is thus also diagonal and easily computed as $\mathrm{diag}(I/\tilde{\alpha}, 1/(1-\tilde{\alpha}))$.
- There sometimes exist analytical relations between $\hat{Q}$ and M; e.g., if $\hat{Q} := [\sin t, \cos t; -\cos t, \sin t]$ varies with time t, then $M = \hat{Q}^{-1} = \hat{Q}^{\mathrm{T}} = [\sin t, -\cos t; \cos t, \sin t]$ [24].

So, in many situations, M can easily be computed, and there shall be no bottleneck for the online computation of M in the dual-neural-network design.

4.2.3 The Third Step—Standard DNN Model

Third, based on solving (4.9), we could thus have the following dynamic equation and output equation of the standard dual neural network to solve QP (4.2)–(4.5) in

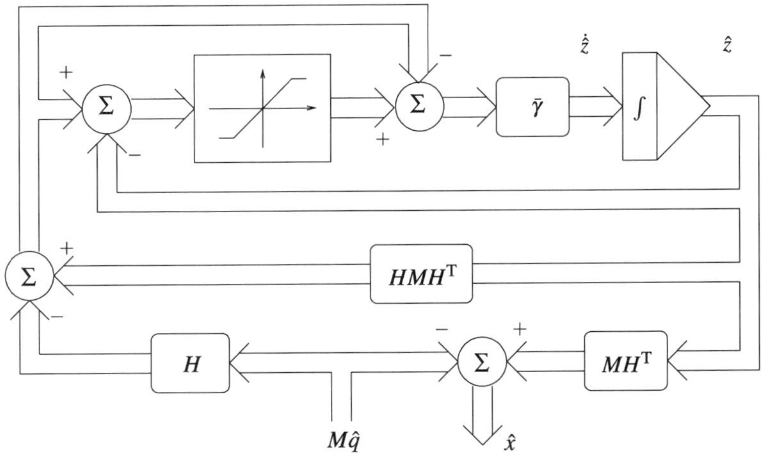

Fig. 4.2 Block diagram of dual neural network (4.10) for solving quadratic program (4.2)–(4.5)

real time [7, 9]:

$$\begin{aligned}\dot{\hat{z}} &= \bar{\gamma}\{\mathscr{P}_\Omega(HMH^\mathrm{T}\hat{z} - HM\hat{q} - \hat{z}) - (HMH^\mathrm{T}\hat{z} - HM\hat{q})\},\\ \hat{x} &= MH^\mathrm{T}\hat{z} - M\hat{q},\end{aligned} \tag{4.10}$$

where $\bar{\gamma} \in R$ is a positive design parameter used to scale the convergence rate of the dual neural network. For superior online performance, the parameter $\bar{\gamma}$, being the reciprocal of a capacitive parameter, is set as large as the hardware would permit [e.g., in analog circuits or very large scale integration (VLSI)] [25]. For simulative and/or experimental purposes, $\bar{\gamma}$ is usually selected between 10^3 and 10^8.

Evidently, with output $\hat{x} \in R^n$, the dynamic equation described in (4.10) shows that

- the dual neural network is composed of only one layer of $m + \dim(\hat{b}) + n$ neurons (in terms of the dimension of state $\hat{z}$) and without using any analog multiplier or penalty parameter, as opposed to the penalty-based neural network approach [1, 2];
- compared to the primal–dual neural network [5, 6], the dynamic equation of dual neural network is piecewise linear without any high-order nonlinear term;
- compared to the early dual-neural-network model [26], the dynamic equation of the standard dual neural network (4.10) is simpler and without any scaling matrix [e.g., there is a scaling matrix $(I + A\hat{Q}^{-1}A^\mathrm{T})$ in the early DNN model];
- to construct the dual neural network (4.10), we need introduce the augmentation and definition of ζ^-, ζ^+, H, $\mathscr{P}_\Omega(\cdot)$, and M.

Consequently, the architecture of the dual neural network to be implemented finally on VLSI is simpler than those of most existing recurrent neural network approaches. The block diagram of the dual neural network (4.10) is depicted in Fig. 4.2. A circuit realizing the dual neural network consists of summers, integrators, and weighted connections. The piecewise-linear activation-function (or activation-function array) $\mathscr{P}_\Omega(\cdot)$ may be implemented by using operational amplifiers known as limiters [1, 8, 10–12, 15, 17, 18, 20, 24, 25].

Remark 4.5 This dual neural network, designed based on KKT conditions, has global (exponential) convergence to optimal solutions of strictly convex QPs (to be shown in the next section). The DNN is of piecewise linear dynamics and capable of handling hybrid constraints (4.3)–(4.5). Note that such a DNN entails the inverse of coefficient matrix, $\hat{Q}^{-1}$, denoted by M in (4.10). In the cases of M being diagonal or analytical (as discussed at the end of Sect. 4.2.2), there shall be no bottleneck in computing M for the DNN.

4.2.4 Inverse-free Dual-Neural-Network Model

For the inverse M being analytical, the standard dual-neural-network model is depicted in (4.10). However, if the inverse M is nondiagonal, nonanalytical, and/or time-varying, there may exist an online matrix-inversion subtask for the dual-neural-network computation involved in (4.10) [7, 15, 24]. This concern introduces the following inverse-free dual-neural-network model.

As inspired by [27], we know that solving for an inverse-related vector-valued term, such as $\hat{Q}^{-1}\hat{q}$, is generally more efficient and economical than solving for the inverse $\hat{Q}^{-1}$ itself. Looking at Theorem 4.2, Eqs. (4.9) and (4.10), we can see that, instead of solving $\hat{Q}^{-1}$ explicitly, we only need solve a system of linear equations, $\hat{Q}\hat{x} = (H^{\mathrm{T}}\hat{z} - \hat{q})$, for $\hat{x}$ [theoretically, $\hat{x} = \hat{Q}^{-1}(H^{\mathrm{T}}\hat{z} - \hat{q}) = M(H^{\mathrm{T}}\hat{z} - \hat{q})$].

To solve $\hat{Q}\hat{x} = (H^{\mathrm{T}}\hat{z} - \hat{q})$ for $\hat{x}$, we can first define a scalar-valued energy function such as $\varepsilon(t) = \|\hat{Q}\hat{x} - (H^{\mathrm{T}}\hat{z} - \hat{q})\|^2/2$ over time t. Then, we use the negative of the gradient $\partial\varepsilon/\partial\hat{x} = \hat{Q}^{\mathrm{T}}(\hat{Q}\hat{x} - (H^{\mathrm{T}}\hat{z} - \hat{q}))$ as the descent direction. Thus, we have a direct $\hat{x}$-solving primal neural network (PNN):

$$
\begin{aligned}
\dot{\hat{x}} &= -\bar{\beta}\frac{\partial\varepsilon}{\partial\hat{x}} \\
&= -\bar{\beta}\hat{Q}^{\mathrm{T}}\big(\hat{Q}\hat{x} - \big(H^{\mathrm{T}}\hat{z} - \hat{q}\big)\big) \\
&= -\bar{\beta}\hat{Q}^{\mathrm{T}}\hat{Q}\hat{x} + \bar{\beta}\hat{Q}^{\mathrm{T}}H^{\mathrm{T}}\hat{z} - \bar{\beta}\hat{Q}^{\mathrm{T}}\hat{q}
\end{aligned}
$$

or simply,

$$\dot{\hat{x}} = -\bar{\beta}\hat{Q}^{\mathrm{T}}\hat{Q}\hat{x} + \bar{\beta}\hat{Q}^{\mathrm{T}}H^{\mathrm{T}}\hat{z} - \bar{\beta}\hat{Q}^{\mathrm{T}}\hat{q}, \tag{4.11}$$

where initial state $\hat{x}(0) \in R^n$; and likewise, the design parameter $\bar{\beta} > 0$ is used to adjust the network-convergence rate of (4.11). Being the reciprocal of a capacitance parameter, $\bar{\beta}$ is set as large as the hardware would permit, or selected appropriately for simulations and/or experiments.

According to the naming of recurrent neural networks in Sect. 4.1.1, for the $\hat{x}$-solving neural network (4.11) alone, it is a primal neural network. This is because the dynamic equation and implementation of such a neural network only use the original/primal decision vector $\hat{x}$, with $\hat{z}$ being an input vector. Furthermore, we have the following theorem on the solution convergence of primal neural network (4.11) alone.

Theorem 4.3 *Assume that matrix $\hat{Q} \in R^{n\times n}$ is nonsingular (note that the positive-definiteness is a special case of matrix nonsingularity). Starting from any initial state $\hat{x}(0) \in R^n$, the $\hat{x}$-solving primal neural network (4.11) is exponentially convergent to the theoretical solution, $\hat{Q}^{-1}(H^{\mathrm{T}}\hat{z} - \hat{q})$. In addition, the global convergence rate λ is the product of $\bar{\beta}$ and the minimal eigenvalue of $\hat{Q}^{\mathrm{T}}\hat{Q}$.*

Proof It can be generalized from [28–30]. □

Remark 4.6 Refer to Definitions 4.2 and 4.7: the global exponential convergence of recurrent neural networks implies an arbitrarily fast convergence to solutions. By the above theorem, the global exponential convergence of primal neural network (4.11) means that, after a time period of $4/\bar{\beta}$, the computational error $\|\hat{x}(t) - \hat{Q}^{-1}(H^{\mathrm{T}}\hat{z} - \hat{q})\|$ will be less than 1.85 % [i.e., $\approx \exp(-4)$] of $\eta\|\hat{x}(0) - \hat{Q}^{-1}(H^{\mathrm{T}}\hat{z} - \hat{q})\|$ for some constant $\eta > 0$. For example, with $\bar{\beta} = 10^6$,

- if the minimal eigenvalue of $\hat{Q}^{\mathrm{T}}\hat{Q}$ is 0.3, then the convergence time $4/\bar{\beta}$ will be less than 44.5 μs; and
- if the minimal eigenvalue is more than 1.0, then the convergence time to get the relative error of 1.85η % will be roughly less than 4 μs.

Starting from any $t_0 \geq 0$ and $\hat{x}(t_0) \in R^n$, after some time instant t_1 such that $\forall t \geq t_1$, $\|\hat{x}(t) - \hat{Q}^{-1}(H^{\mathrm{T}}\hat{z} - \hat{q})\| \leq \varepsilon$ for some $\varepsilon > 0$, or directly after time instant $t_1 \geq (4/\bar{\beta} + t_0)$, the output of the $\hat{x}$-solving primal neural network (4.11) [i.e., $\hat{x}(t)$] will be usable as the approximation of true $\hat{x}$-term, $\hat{Q}^{-1}(H^{\mathrm{T}}\hat{z} - \hat{q})$, in the computation of DNN (4.10).

Thus, it follows from the model equations (4.10) and (4.11) that we have the combined inverse-free dual neural network as below:

$$\begin{aligned} \dot{\hat{x}} &= -\bar{\beta}\hat{Q}^{\mathrm{T}}\hat{Q}\hat{x} + \bar{\beta}\hat{Q}^{\mathrm{T}}H^{\mathrm{T}}\hat{z} - \bar{\beta}\hat{Q}^{\mathrm{T}}\hat{q}, \\ \dot{\hat{z}} &= \bar{\gamma}\{\mathscr{P}_{\Omega}(H\hat{x} - \hat{z}) - H\hat{x}\}. \end{aligned} \tag{4.12}$$

- Evidently, after combining the $\hat{x}$-solving primal neural network (4.11) with the standard dual neural network (4.10), the resultant inverse-free dual neural network (4.12) has actually become a new type of primal–dual neural networks. This is in view of Concept 4.3; that is, the dynamic equation and implementation of the combined inverse-free dual neural network (4.12) use a group of dual decision vector $\hat{z}$, in addition to using the original/primal decision vector $\hat{x}$.
- Moreover, the combined inverse-free dual neural network (4.12) is piecewise linear (or simply linear) and has no high-order nonlinear terms or penalty parameters.

- Similar to Remark 4.5, this combined inverse-free dual neural network (4.12) is designed also based on KKT conditions and has global (exponential) convergence to optimal solutions of convex quadratic programs.
- Note that the term $\hat{Q}^{-1}$ (i.e., M) is no more necessary in the combined inverse-free dual neural network (4.12), as compared to the standard dual neural network (4.10).

4.3 Dual-Neural-Network Analysis

In this section, we show the global convergence and exponential convergence of the presented dual neural network (4.10) for solving strictly-convex QP (4.2)–(4.5). Before presenting the main results (Theorems 4.5 through 4.8), it is worth mentioning that some related definitions have already been given in Sect. 4.1.1 and that some related lemma and theorem are presented below.

Lemma 4.1 *Assume that the set $\Omega \subset R^{m+\dim(\hat{b})+n}$ is a closed convex set. Then the following two inequalities hold* [22, 31, 32]:

$$\left(\hat{z} - \mathscr{P}_\Omega(\hat{z})\right)^{\mathrm{T}}\left(\mathscr{P}_\Omega(\hat{z}) - \tilde{z}\right) \geq 0 \quad \forall \hat{z} \in R^{m+\dim(\hat{b})+n}, \tilde{z} \in \Omega;$$

$$\left\|\mathscr{P}_\Omega(\hat{z}) - \mathscr{P}_\Omega(\tilde{z})\right\| \leq \|\hat{z} - \tilde{z}\| \quad \forall \hat{z}, \tilde{z} \in R^{m+\dim(\hat{b})+n};$$

where $\mathscr{P}_\Omega(\hat{z}) = \arg\min_{\tilde{z}\in\Omega} \|\hat{z} - \tilde{z}\|$ *is a projection operator from* $R^{m+\dim(\hat{b})+n}$ *to* Ω.

It is evident that the projection set $\Omega = \{\hat{z} \in R^{m+\dim(\hat{b})+n} | \zeta^- \leq \hat{z} \leq \zeta^+\}$, defined in Sect. 4.2 for dual neural network, is such a closed convex set satisfying the above lemma, and the projection operator $\mathscr{P}_\Omega(\cdot)$ possesses the above property.

Theorem 4.4 *At the equilibrium point $\hat{z}^*$ of dual neural network* (4.10) *that satisfies $\mathscr{P}_\Omega(HMH^{\mathrm{T}}\hat{z}^* - HM\hat{q} - \hat{z}^*) - (HMH^{\mathrm{T}}\hat{z}^* - HM\hat{q}) = 0$, we have the following inequality property*:

$$\left(\tilde{z} - HMH^{\mathrm{T}}\hat{z}^* + HM\hat{q}\right)^{\mathrm{T}}\hat{z}^* \geq 0 \quad \forall \tilde{z} \in \Omega. \tag{4.13}$$

Proof It can be generalized from [23]. First, from the output equation of dual neural network (4.10), $\hat{x} = MH^{\mathrm{T}}\hat{z} - M\hat{q}$, we know that $HMH^{\mathrm{T}}\hat{z}^* - HM\hat{q} = H\hat{x}^*$, where $\hat{x}^*$ is the optimal solution to QP (4.2)–(4.5). This is in view of the satisfaction of Eq. (4.9) and Theorem 4.2. Then, we discuss the following three cases.

- Case 1. If for some $i \in \{1, 2, \ldots, (m + \dim(\hat{b}) + n)\}$, $\hat{z}_i^* = 0$ and $\zeta_i^- \leq [H\hat{x}^*]_i \leq \zeta_i^+$, then $(\tilde{z}_i - [H\hat{x}^*]_i)\hat{z}_i^* = 0$;
- Case 2. If for some $j \in \{1, 2, \ldots, (m + \dim(\hat{b}) + n)\}$, $\hat{z}_j^* > 0$, $[H\hat{x}^*]_j = \zeta_j^-$, and $\zeta_j^- \leq \tilde{z}_j \leq \zeta_j^+$, then $\tilde{z}_j - [H\hat{x}^*]_j \geq 0$, and thus $(\tilde{z}_j - [H\hat{x}^*]_j)\hat{z}_j^* \geq 0$; and

- Case 3. If for some $k \in \{1, 2, \ldots, (m + \dim(\hat{b}) + n)\}$, $\hat{z}_k^* < 0$, $[H\hat{x}^*]_k = \zeta_k^+$, and $\zeta_k^- \le \tilde{z}_k \le \zeta_k^+$, then $\tilde{z}_k - [H\hat{x}^*]_k \le 0$, and thus $(\tilde{z}_k - [H\hat{x}^*]_k)\hat{z}_k^* \ge 0$.

Therefore, by combining these three cases (using index i to represent i, j, and k) we have

$$\left(\tilde{z} - HMH^{\mathrm{T}}\hat{z}^* + HM\hat{q}\right)^{\mathrm{T}}\hat{z}^* = \sum_{i=1}^{(m+\dim(\hat{b})+n)} \left(\tilde{z}_i - \left[H\hat{x}^*\right]_i\right)\hat{z}_i^* \ge 0 \quad \forall \tilde{z} \in \Omega,$$

which completes the proof. □

4.3.1 Global Convergence

Now, assuming the existence of optimal solution $\hat{x}^*$ to QP (4.2)–(4.5), we have the following theorems about the convergence of dual neural network.

Theorem 4.5 *Starting from any initial state* $\hat{z}(0) \in R^{m+\dim(\hat{b})+n}$, *the dual neural network* (4.10) *converges to an equilibrium point* $\hat{z}^*$.

Proof To show the convergence property, the following inequalities, especially Eq. (4.17), are derived. First, according to Theorem 4.4, at $\hat{z}^*$, we have the inequality property (4.13), i.e., $(\tilde{z} - HMH^{\mathrm{T}}\hat{z}^* + HM\hat{q})^{\mathrm{T}}\hat{z}^* \ge 0\ \forall \tilde{z} \in \Omega$. Therefore, it follows that

$$\left(\mathscr{P}_\Omega\left(HMH^{\mathrm{T}}\hat{z} - HM\hat{q} - \hat{z}\right) - HMH^{\mathrm{T}}\hat{z}^* + HM\hat{q}\right)^{\mathrm{T}}\hat{z}^* \ge 0. \tag{4.14}$$

Second, by defining $\tilde{\hat{z}} := HMH^{\mathrm{T}}\hat{z} - HM\hat{q}$ it follows from Lemma 4.1 that, for all $\hat{z} \in R^{m+\dim(\hat{b})+n}$,

$$\left(\mathscr{P}_\Omega(\tilde{\hat{z}} - \hat{z}) - HMH^{\mathrm{T}}\hat{z}^* + HM\hat{q}\right)^{\mathrm{T}}\left(\tilde{\hat{z}} - \hat{z} - \mathscr{P}_\Omega(\tilde{\hat{z}} - \hat{z})\right) \ge 0. \tag{4.15}$$

Third, adding Eq. (4.14) and (4.15) yields

$$\left(\mathscr{P}_\Omega(\tilde{\hat{z}} - \hat{z}) - HMH^{\mathrm{T}}\hat{z}^* + HM\hat{q}\right)^{\mathrm{T}}\left(\hat{z}^* + \tilde{\hat{z}} - \hat{z} - \mathscr{P}_\Omega(\tilde{\hat{z}} - \hat{z})\right) \ge 0. \tag{4.16}$$

Defining $\tilde{g} = \mathscr{P}_\Omega(HMH^{\mathrm{T}}\hat{z} - HM\hat{q} - \hat{z}) - HMH^{\mathrm{T}}\hat{z} + HM\hat{q}$, i.e., $\tilde{g} = \mathscr{P}_\Omega(\tilde{\hat{z}} - \hat{z}) - \tilde{\hat{z}}$, Eq. (4.16) is thus reformulated as

$$\left(\tilde{g} + HMH^{\mathrm{T}}(\hat{z} - \hat{z}^*)\right)^{\mathrm{T}}\left((\hat{z} - \hat{z}^*) + \tilde{g}\right) \le 0$$

and, subsequently, as

$$(\hat{z} - \hat{z}^*)^{\mathrm{T}}\tilde{g} + \tilde{g}^{\mathrm{T}}HMH^{\mathrm{T}}(\hat{z} - \hat{z}^*) \le -\|\tilde{g}\|^2 - (\hat{z} - \hat{z}^*)^{\mathrm{T}}HMH^{\mathrm{T}}(\hat{z} - \hat{z}^*). \tag{4.17}$$

Now we choose a Lyapunov function candidate as

$$V(\hat{z}(t)) = \frac{1}{2}\|K(\hat{z}(t) - \hat{z}^*)\|^2, \tag{4.18}$$

where the matrix K is symmetric positive-definite, and $K^2 = (I + HMH^T)/\bar{\gamma} > 0$. Evidently, $V(\hat{z})$ is positive-definite (i.e., $V > 0$ if $\hat{z} \neq \hat{z}^*$, and $V = 0$ if and only if $\hat{z} = \hat{z}^*$) for any $\hat{z}$ taken in the domain $\Gamma(\hat{z}^*) \subset R^{m+\dim(\hat{b})+n}$ (i.e., the attraction region of $\hat{z}^*$). dV/dt is negative definite, since, in view of (4.17),

$$\begin{aligned}
\frac{dV}{dt} &= (\hat{z} - \hat{z}^*)^T K^2 \dot{\hat{z}} \\
&= (\hat{z} - \hat{z}^*)^T (I + HMH^T)\tilde{g} \\
&\leq -\|\tilde{g}\|^2 - (\hat{z} - \hat{z}^*)^T HMH^T (\hat{z} - \hat{z}^*) \leq 0,
\end{aligned} \tag{4.19}$$

and $dV/dt = 0$ if and only if $\hat{z} = \hat{z}^*$ in $\Gamma(\hat{z}^*)$. Thus, it follows that the dual neural network (4.10) is globally convergent to an equilibrium point $\hat{z}^*$, which depends on the initial state $\hat{z}(0) \in R^{m+\dim(\hat{b})+n}$ of the trajectory $\hat{z}(t)$. For a better understanding on the global convergence, please also refer to Definition 4.1 and Remark 4.1 as discussed in Sect. 4.1.1. □

Theorem 4.6 *At the equilibrium point* $\hat{z}^*$ *of the dual neural network* (4.10), *the network output* $\hat{x}^* = MH^T\hat{z}^* - M\hat{q}$ *is the optimal solution to QP* (4.2)–(4.5).

Proof The equilibrium point $\hat{z}^*$ of the dual neural network (4.10) satisfies the following equations:

$$\begin{aligned}
&\mathscr{P}_\Omega(HMH^T\hat{z}^* - HM\hat{q} - \hat{z}^*) - HMH^T\hat{z}^* + HM\hat{q} = 0, \\
&\hat{x}^* = MH^T\hat{z}^* - M\hat{q},
\end{aligned}$$

which is equivalent to

$$\mathscr{P}_\Omega(H\hat{x}^* - \hat{z}^*) - H\hat{x}^* = 0. \tag{4.20}$$

Let u^*, v^*, and ν^* denote the vectors of dual decision vectors corresponding to the constraints (4.3), (4.4), and (4.5), respectively. That is, $\hat{z}^* = [(u^*)^T, (v^*)^T, (\nu^*)^T]^T \in R^{m+\dim(\hat{b})+n}$. Equation (4.20) becomes

$$\mathscr{P}_\Omega\left(\begin{bmatrix} J\hat{x}^* - u^* \\ A\hat{x}^* - v^* \\ \hat{x}^* - \nu^* \end{bmatrix}\right) - \begin{bmatrix} J\hat{x}^* \\ A\hat{x}^* \\ \hat{x}^* \end{bmatrix} = 0. \tag{4.21}$$

By the definition of $\mathscr{P}_\Omega(\cdot)$ given immediately before Theorem 4.2, it follows from (4.21) that $\hat{d} - J\hat{x}^* = 0$ with u^* unrestricted and

$$\begin{cases} A\hat{x}^* - \hat{b} \le 0, \quad v^* \le 0, \\ (v^*)^{\mathrm{T}}(A\hat{x}^* - \hat{b}) = 0, \end{cases} \qquad \begin{cases} \hat{x}_i^* = \xi_i^- & \text{if } \nu_i^* > 0, \\ \hat{x}_i^* = \xi_i^+ & \text{if } \nu_i^* < 0, \\ \xi_i^- \le \hat{x}_i^* \le \xi_i^+ & \text{if } \nu_i^* = 0. \end{cases}$$

In addition to $\hat{Q}\hat{x}^* - J^{\mathrm{T}}u^* - A^{\mathrm{T}}v^* - \nu^* + \hat{q} = 0$ implied by $\hat{x}^* = MH^{\mathrm{T}}\hat{z}^* - M\hat{q}$, the above equations constitute the KKT optimality conditions of (4.2)–(4.5). The proof is thus completed. It is worth noting that, for a better understanding on the KKT conditions, please also refer to Sect. 4.1.2 about the "Primal and Dual Problems" and Theorem 4.2. □

Remark 4.7 Theorems 4.5 and 4.6 actually imply that the network output $\hat{x}(t) = MH^{\mathrm{T}}\hat{z}(t) - M\hat{q}$ of the dual neural network (4.10) is globally convergent to the unique optimal solution $\hat{x}^* = MH^{\mathrm{T}}\hat{z}^* - M\hat{q}$ of strictly-convex QP (4.2)–(4.5). The reason why $\hat{x}^*$ is unique is that the coefficient matrix $\hat{Q}$ in QP (4.2)–(4.5) is assumed in this chapter to be positive definite [19]. If $\hat{Q}$ is positive semi-definite (or zero, i.e., the LP case), $\hat{x}^*$ is not unique, and there may exist an infinite number of nonisolated solutions denoted by $\hat{x}^*$.

4.3.2 Exponential Convergence

Now we come to prove, theoretically, under what conditions or assumptions the dual neural network (4.10) is globally exponentially convergent to an equilibrium point $\hat{z}^*$, with the network output $\hat{x}^* = MH^{\mathrm{T}}\hat{z}^* - M\hat{q}$ constituting the optimal solution to QP (4.2)–(4.5). Based on the existence of optimal solution $\hat{x}^*$, the following theorems are presented.

Theorem 4.7 *Starting from any initial state $\hat{z}(0) \in R^{m+\dim(\hat{b})+n}$, the dual neural network* (4.10) *is globally exponentially convergent to an equilibrium point $\hat{z}^*$, provided that there exists a constant $\varrho > 0$ such that*

$$\left\| \mathscr{P}_\Omega\left(HMH^{\mathrm{T}}\hat{z} - \hat{z} - HM\hat{q}\right) - HMH^{\mathrm{T}}\hat{z} + HM\hat{q} \right\|^2 \ge \varrho \left\| \hat{z} - \hat{z}^* \right\|^2. \qquad (4.22)$$

In addition, the convergence rate is proportional to $\bar{\gamma}\varrho$.

Proof First, we have Theorem 4.5 and its proof. To show the global exponential convergence, we review $V(\hat{z})$ and $\mathrm{d}V/\mathrm{d}t$ that were defined at the end of the proof of Theorem 4.5. That is, the Lyapunov function candidate $V(\hat{z}(t)) :=$

$\|K(\hat{z}(t)-\hat{z}^*)\|^2/2$ with matrix $K^2=(I+HMH^{\mathrm{T}})/\bar{\gamma}>0$. Evidently, $V(\hat{z})$ is positive definite, and $\mathrm{d}V/\mathrm{d}t$ is negative definite. In addition,

$$\frac{\mathrm{d}V}{\mathrm{d}t}=(\hat{z}-\hat{z}^*)^{\mathrm{T}}K^2\dot{\hat{z}}=(\hat{z}-\hat{z}^*)^{\mathrm{T}}(I+HMH^{\mathrm{T}})\tilde{g}$$
$$\leq -\|\tilde{g}\|^2-(\hat{z}-\hat{z}^*)^{\mathrm{T}}HMH^{\mathrm{T}}(\hat{z}-\hat{z}^*)\leq 0,$$

which shows that the dual neural network (4.10) is globally convergent to an equilibrium point $\hat{z}^*$, which depends on the initial state $\hat{z}(0)\in R^{m+\dim(\hat{b})+n}$ of the trajectory $\hat{z}(t)$.

It follows from the definition of $V(\hat{z}(t))$ that $\hat{c}_1\|\hat{z}-\hat{z}^*\|^2\leq V(\hat{z})\leq \hat{c}_2\|\hat{z}-\hat{z}^*\|^2$, where the constants $\hat{c}_2\geq\hat{c}_1>0$ are respectively the maximal and minimal eigenvalues of $(I+HMH^{\mathrm{T}})/\bar{\gamma}$. Clearly, $\hat{c}_1$ and $\hat{c}_2$ are proportional to the reciprocal of the design parameter $\bar{\gamma}$ (i.e., $1/\bar{\gamma}$). Moreover, we have the exponential convergence condition (ECC) depicted in (4.22); that is, there exists a constant $\varrho>0$ such that $\|\mathscr{P}_\Omega(HMH^{\mathrm{T}}\hat{z}-\hat{z}-HM\hat{q})-HMH^{\mathrm{T}}\hat{z}+HM\hat{q}\|^2\geq\varrho\|\hat{z}-\hat{z}^*\|^2$. In addition, review Eq. (4.17); that is, $(\hat{z}-\hat{z}^*)^{\mathrm{T}}\tilde{g}+\tilde{g}^{\mathrm{T}}HMH^{\mathrm{T}}(\hat{z}-\hat{z}^*)\leq-\|\tilde{g}\|^2-(\hat{z}-\hat{z}^*)^{\mathrm{T}}HMH^{\mathrm{T}}(\hat{z}-\hat{z}^*)$. We thus have

$$\frac{\mathrm{d}V(\hat{z})}{\mathrm{d}t}\leq-\varrho\|\hat{z}-\hat{z}^*\|^2-(\hat{z}-\hat{z}^*)^{\mathrm{T}}HMH^{\mathrm{T}}(\hat{z}-\hat{z}^*)$$
$$=-(\hat{z}-\hat{z}^*)^{\mathrm{T}}(\varrho I+HMH^{\mathrm{T}})(\hat{z}-\hat{z}^*)\leq-\bar{\alpha}V(\hat{z}(t)),$$

where the constant $\bar{\alpha}=\varrho/\hat{c}_2>0$. Thus, we have $V(\hat{z}(t))=O(\exp(-\bar{\alpha}(t-t_0)))$ $\forall t\geq t_0\geq 0$, and hence $\|\hat{z}(t)-\hat{z}^*\|=O(\exp(-\bar{\alpha}(t-t_0)/2))$ $\forall t\geq t_0\geq 0$, where the exponential convergence rate $\bar{\alpha}$ is proportional to $\bar{\gamma}\varrho$ because $1/\hat{c}_2$ is proportional to $\bar{\gamma}$. □

Remark 4.8 The exponential convergence condition (ECC) depicted in (4.22) is actually a reasonably mild condition (or assumption). It follows from the proof of Theorem 4.5 that $\|\mathscr{P}_\Omega(HMH^{\mathrm{T}}\hat{z}-\hat{z}-HM\hat{q})-HMH^{\mathrm{T}}\hat{z}+HM\hat{q}\|^2=\|\tilde{g}\|^2$, with $\tilde{g}$ defined between Eqs. (4.16) and (4.17). This exponential convergence condition (4.22) then becomes that there is $\varrho>0$ such that $\|\tilde{g}\|^2\geq\varrho\|\hat{z}-\hat{z}^*\|^2$. It is reasonable, because $\|\tilde{g}\|>0$ for any $\hat{z}(t)\neq\hat{z}^*$ in $\Gamma(\hat{z}^*)$ and $\|\tilde{g}\|=0$ amounts to $\hat{z}=\hat{z}^*$. In addition, by analyzing the linear/saturation cases of $\mathscr{P}_\Omega([HMH^{\mathrm{T}}\hat{z}-HM\hat{q}-\hat{z}]_j)$, the range of ϱ is (0, 1]. Numerical results also show that such a condition does exist in practice, though it is hard to prove.

Remark 4.9 The proof complexity of the above ECC is investigated by analyzing its one-dimensional case [26]. It is shown that such a one-dimensional case always holds. Specifically, all possible cases (including subcases) of the one-dimensional form of the original ECC (4.22) are discussed in [26] with about 6 pages. By applying the proved convergence property of the dual neural network (i.e., "$\hat{z}(t)\to\hat{z}^*$ as $t\to\infty$" summarized in Theorem 4.5), it is shown that there always exists a lower bound $\varrho>0$ such that condition $|\mathscr{P}_\Omega(\tilde{m}\hat{z}-\hat{z})-\tilde{m}\hat{z}|^2\geq\varrho|\hat{z}-\hat{z}^*|^2$ always holds, provided that the initial state $\hat{z}(0)$ is not selected at $\pm\infty$, where $HMH^{\mathrm{T}} := \tilde{m}$ for simplicity, and $\hat{q} = 0$ is assumed. In addition, the basic tools in the proof are equilibrium equation $\mathscr{P}_{\Omega}(\tilde{m}\hat{z}^* - \hat{z}^*) - \tilde{m}\hat{z}^* = 0$ and the piecewise linearity of $\mathscr{P}_{\Omega}(\cdot)$. The proof results are also consistent with the analysis of ECC (4.22) in Remark 4.8.

Theorem 4.8 *The network output* $\hat{x}(t)$ *of the dual neural network* (4.10) *is globally exponentially convergent to the unique optimal solution* $\hat{x}^*$ *of strictly convex QP* (4.2)–(4.5), *provided that the exponential convergence condition* (*ECC*) (4.22) *holds.*

Proof We know that the objective function in (4.2) is strictly convex due to the positive definiteness of the coefficient matrix $\hat{Q}$ and that the constraint region constructed by (4.3)–(4.5) is a convex set. By Theorem 3.4.2 of [19], in view of the above points, we can conclude that the solution $\hat{x}^*$ to quadratic-programming problem (4.2)–(4.5) is unique. Please also refer to Remark 4.7.

It follows from the network-output equation, $\hat{x}(t) = MH^{\mathrm{T}}\hat{z}(t) - M\hat{q}$, and Theorem 4.7 that, for all $\hat{x}(0) \in R^n$ and $\hat{z}(0) \in R^{m+\dim(\hat{b})+n}$,

$$\left\|\hat{x}(t) - \hat{x}^*\right\| = \left\|MH^{\mathrm{T}}\big(\hat{z}(t) - \hat{z}^*\big)\right\| \le \left\|MH^{\mathrm{T}}\right\| \left\|\hat{z}(t) - \hat{z}^*\right\|. \tag{4.23}$$

In view of the results of Theorem 4.7 that $\|\hat{z}(t) - \hat{z}^*\| = O(\exp(-\bar{\alpha}(t-t_0)/2))$ $\forall t \ge t_0 \ge 0$, inequality (4.23) implies that the network output $\hat{x}(t)$ of the dual neural network (4.10) exponentially converges to $\hat{x}^*$, which is the unique solution to the constrained quadratic program (4.2)–(4.5). □

Remark 4.10 To solve QP (4.2)–(4.5) in real time and in an error-free parallel-computing manner, the standard dual-neural-network model (4.10) is proposed. Global exponential stability is one of the most desirable dynamic properties. Being globally exponentially convergent, dual neural networks can converge to their solutions most rapidly. The proof complexity of such an exponential convergence condition (ECC) (4.22) is mentioned by analyzing its one-dimensional case [26]. The conclusions are that the one-dimensional case of ECC always holds and that the ECC proof is quite complex with many subcases. Similar theoretical results could also be achieved for the early dual-neural-network model depicted in [26] and the inverse-free dual-neural-network model (4.12).

4.4 Dual-Neural-Network Simulation

The design and analysis of dual neural networks (4.10) and (4.12) have been presented in the previous sections. In this section, we show some illustrative simulation results related to the dual neural networks.

First, we can see that all the dynamic equations depicting the dual neural networks are differential equations of the form of $\dot{\hat{z}} = f(t, \hat{z})$ with initial condition $t_0 \ge 0$ and given $\hat{z}_0 = \hat{z}(t_0)$. Such a dynamic equation is actually an initial-value ordinary-differential-equation (ODE). Then, to simulate the dual neural network [e.g., the standard model (4.10)], we can use MATLAB ODE routine "ode45" [33] to simulate such an initial-value ODE. The syntax of using "ode45" to simulate the standard DNN model (4.10) is simply that "[t, y]=ode45(@standardDNNmodel 4.10rightSide, [t0, tFinal], y0)", where "standardDNNmodel4.10rightSide" is a function handle returning a column vector corresponding to $f(t,\hat{z})$, i.e., $\bar{\gamma}\{\mathscr{P}_\Omega(HMH^{\mathrm{T}}\hat{z}-HM\hat{q}-\hat{z})-(HMH^{\mathrm{T}}\hat{z}-HM\hat{q})\}$ being the right-hand side of Eq. (4.10). Now we come to the illustrative simulation results.

A classic numerical example is considered here to demonstrate the effectiveness and performance behavior of the presented dual neural network (4.10) for solving QPs subject to general linear constraints [9]:

$$
\begin{aligned}
&\text{minimize } 11\hat{x}_1^2+\hat{x}_2^2+\hat{x}_3^2-2\hat{x}_1\hat{x}_2+6\hat{x}_1\hat{x}_3-4\hat{x}_1,\\
&\text{subject to } 2\hat{x}_1+2\hat{x}_2+\hat{x}_3=0,\\
&\qquad -\hat{x}_1+\hat{x}_2\le -1,\\
&\qquad 3\hat{x}_1+\hat{x}_3\le 4,\\
&\qquad -6\le \hat{x}_1,\hat{x}_2,\hat{x}_3\le 6.
\end{aligned}
$$

That is, the coefficients $J=[2,2,1]$, $\hat{d}=0$, $\xi^+=-\xi^-=[6,6,6]^{\mathrm{T}}$, and

$$
\hat{Q}=\begin{bmatrix}22&-2&6\\-2&2&0\\6&0&2\end{bmatrix},\quad \hat{q}=\begin{bmatrix}-4\\0\\0\end{bmatrix},\quad A=\begin{bmatrix}-1&1&0\\3&0&1\end{bmatrix},\quad \hat{b}=\begin{bmatrix}-1\\4\end{bmatrix}.
$$

The dual neural network with $\bar{\gamma}=10^8$ is simulated by using MATLAB ode45, and the results are illustrated in Figs. 4.3, 4.4, 4.5 and 4.6. As shown in Figs. 4.3 and 4.4, starting with any initial state randomly selected within $[-5,5]^6$, the dual neural network converges to the optimal solution of the constrained quadratic program. That is, within 5×10^{-7} second (namely, 0.5 μs), the network can approach $\hat{x}^*=[2,1,-6]^{\mathrm{T}}$ and $\hat{z}^*=[0,-2,0,0,0,0]^{\mathrm{T}}$ without any appreciable error. In addition, the difference between the DNN-computed solution and the theoretical solution $\hat{x}^*$ is less than 10^{-12}. Figure 4.5 illustrates the transient of $\|\tilde{g}\|/\|\hat{z}-\hat{z}^*\|$ during solving the quadratic program. Hence, the EEC constant $\varrho>0$, appearing in Theorem 4.7 and Remark 4.8, can be chosen as the lower bound of $\|\tilde{g}\|/\|\hat{z}-\hat{z}^*\|$, which is about 0.15 in the example. The nonzero ϱ guarantees the exponential convergence property of the dual neural network. This result is also consistent with the analysis of the exponential convergence condition given in Remarks 4.8 and 4.9. As illustrated in Fig. 4.6, the outputs of the dual neural network, starting from five different initial conditions, all converge to the optimal solution of the minimization problem. This substantiates the global convergence property of the dual neural network (4.10).

Before ending this section, it is worth mentioning the following comparisons. If using the Lagrangian neural network [3] to solve the above quadratic program, the total number of neurons in the simulation/implementation is 20, in addition to the premature defect [4]. If using the primal–dual neural networks or other kind of dual

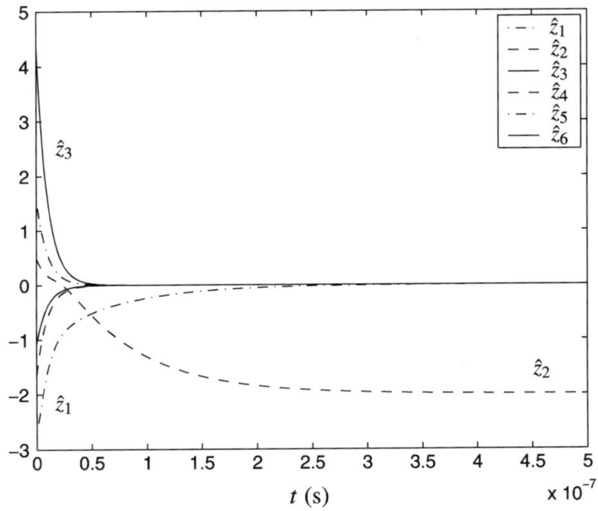

Fig. 4.3 State transients $\hat{z}(t)$ of dual neural network (4.10) for solving a quadratic program

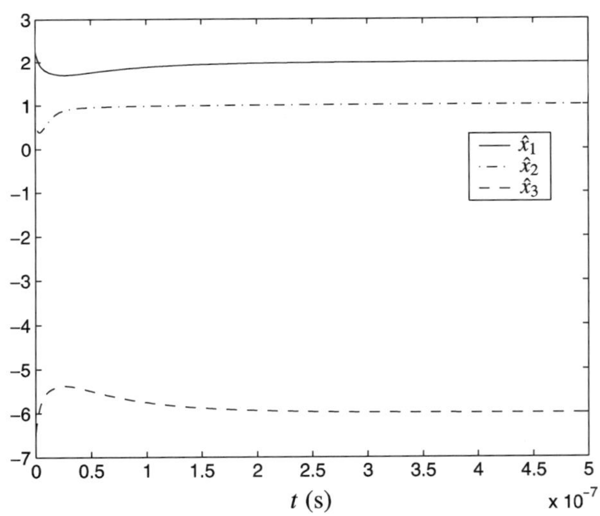

Fig. 4.4 Output transients $\hat{x}(t)$ of dual neural network (4.10) for solving a quadratic program

networks [5, 6, 20], the total number of neurons in the simulation/implementation is usually more than 10. In contrast, the size of the presented standard dual neural network (4.10) to solve the above quadratic-programming problem is only 6. As compared to primal–dual neural networks or other kind of dual neural networks [5, 6, 20], the simulation/implementation complexity reduction of the presented standard dual neural network is more than $(10-6)/6 \approx 66\,\%$.

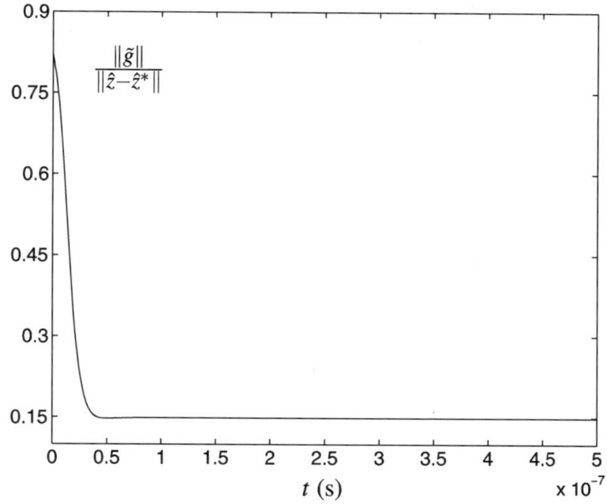

Fig. 4.5 Transient of exponential convergence condition (4.22) of dual neural network (4.10) in solving a quadratic program

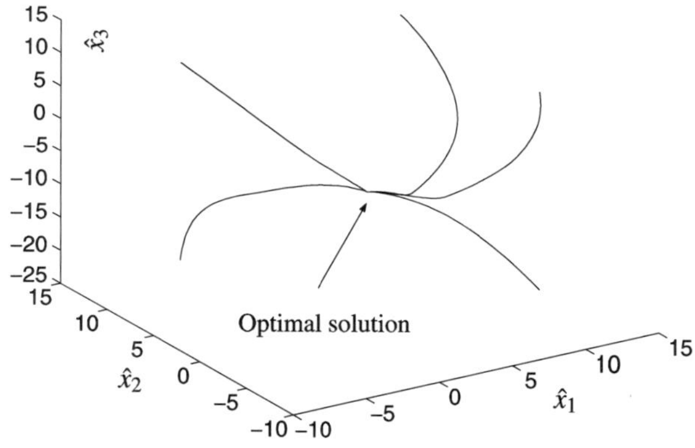

Fig. 4.6 Spatial trajectories of output $(\hat{x}_1, \hat{x}_2, \hat{x}_3)$ of the dual neural network starting from five different initial conditions

4.5 DNN for Solving RMP Scheme

In this section, we tailor and present a dual-neural-network-based QP solver for the real-time drift-free redundancy resolution of physically constrained manipulators. Its design procedure is presented as follows.

First, the quadratic program (2.22)–(2.24) can be rewritten as

$$\begin{aligned} &\text{minimize } \frac{1}{2}\hat{x}^{\mathrm{T}}Q\hat{x} + \hat{q}^{\mathrm{T}}\hat{x}, \\ &\text{subject to } \zeta^- \le H\hat{x} \le \zeta^+, \end{aligned} \tag{4.24}$$

where $\zeta^- := [\hat{d}^{\mathrm{T}}, (\xi^-)^{\mathrm{T}}]^{\mathrm{T}}$, $\zeta^+ := [\hat{d}^{\mathrm{T}}, (\xi^+)^{\mathrm{T}}]^{\mathrm{T}}$ in this dual-neural-network design, and, with I denoting the identity matrix, the matrix H is defined here as

$$H = \begin{bmatrix} J \\ I \end{bmatrix} \in R^{(n+m)\times n}.$$

Second, at time instant t, the QP problem depicted in (2.22)–(2.24) could be viewed as a parametric optimization problem. Now, let us analyze the optimal solution to it and its equivalent problem formulation (4.24). By the KKT conditions [19], we know that $\hat{x}$ is a solution to (4.24) if and only if there is a dual decision variable vector $\hat{z} \in R^{m+n}$ such that $\hat{x} - H^{\mathrm{T}}\hat{z} + \hat{q} = 0$ [in view of $Q = I = Q^{-1}$ in the most preferred RMP scheme (2.2)–(2.5)] and

$$\begin{cases} [H\hat{x}]_i = \zeta_i^- & \text{if } \hat{z}_i > 0, \\ \zeta_i^- \le [H\hat{x}]_i \le \zeta_i^+ & \text{if } \hat{z}_i = 0, \\ [H\hat{x}]_i = \zeta_i^+ & \text{if } \hat{z}_i < 0. \end{cases} \tag{4.25}$$

In addition, condition (4.25) is equivalent to the system of piecewise linear equations $H\hat{x} = \mathscr{P}_\Omega(H\hat{x} - \hat{z})$ (see [9] and references therein), where $\mathscr{P}_\Omega(\cdot)$ is a projection operator from R^{n+m} onto $\Omega := \{\hat{z} | \zeta^- \le \hat{z} \le \zeta^+\} \subset R^{n+m}$, and the ith output of $\mathscr{P}_\Omega(\hat{z})$ is defined as

$$[\mathscr{P}_\Omega(\hat{z})]_i = \begin{cases} \zeta_i^- & \text{if } \hat{z}_i < \zeta_i^-, \\ \hat{z}_i & \text{if } \zeta_i^- \le \hat{z}_i \le \zeta_i^+, \\ \zeta_i^+ & \text{if } \hat{z}_i > \zeta_i^+, \end{cases} \quad i = 1, 2, \ldots, n+m.$$

Third, by the above analysis, a necessary and sufficient condition for solving QP (4.24) is that $\hat{x}$ and $\hat{z}$ satisfy $\hat{x} - H^{\mathrm{T}}\hat{z} + \hat{q} = 0$ and $H\hat{x} = \mathscr{P}_\Omega(H\hat{x} - \hat{z})$. Substituting the former equation into the latter, we have

$$\begin{cases} \hat{x} = H^{\mathrm{T}}\hat{z} - \hat{q}, \\ \mathscr{P}_\Omega(HH^{\mathrm{T}}\hat{z} - H\hat{q} - \hat{z}) = HH^{\mathrm{T}}\hat{z} - H\hat{q}, \end{cases}$$

which gives rise to the dynamic equation of dual neural network solving QP (4.24) as well as QP (2.22)–(2.24):

$$\dot{\hat{z}} = \bar{\gamma}\big(\mathscr{P}_\Omega\big(HH^{\mathrm{T}}\hat{z} - H\hat{q} - \hat{z}\big) - HH^{\mathrm{T}}\hat{z} + H\hat{q}\big), \tag{4.26}$$

$$\hat{x} = H^{\mathrm{T}}\hat{z} - \hat{q}, \tag{4.27}$$

where $\bar{\gamma} > 0$, as before, is a design parameter used to scale the convergence rate of neural networks. Furthermore, we have the following important results [9].

Theorem 4.9 *Assume the existence of optimal solution $\hat{x}^*$ to strictly-convex QP problem* (2.22)–(2.24) *and* (4.24) (*in view of $Q = I$*). *Starting from any initial state $\hat{z}(0)$, dual neural network* (4.26) *converges to an equilibrium point $\hat{z}^*$, of which the* *neural-network output* $\hat{x}^* = H^{\mathrm{T}}\hat{z}^* - \hat{q}$ *is the optimal solution to QP* (4.24) *as well as to the inverse-kinematics problem* (2.22)–(2.24). *Moreover, if there exists a constant* $\rho > 0$ *such that* $\|\mathscr{P}_{\Omega}(HH^{\mathrm{T}}\hat{z} - H\hat{q} - \hat{z}) - (HH^{\mathrm{T}}\hat{z} - H\hat{q})\|_2^2 \geq \rho\|\hat{z} - \hat{z}^*\|_2^2$, *then the exponential convergence can be achieved for such a dual neural network. The exponential convergence rate is proportional to* $\bar{\gamma}\rho$.

4.6 Chapter Conclusions

In this chapter, we have investigated almost every aspect of the DNN, which is one of state-of-the-art recurrent neural networks for solving QPs in real time. The dual neural network is of simple piecewise-linear dynamics and has global (exponential) convergence to optimal solutions. We have also introduced some related concepts and definitions as the preliminaries of this RNN study. Regarding the theoretical aspects of this study, we have presented the analysis results of dual neural networks and their exponential convergence condition. For future application, we have presented the simulation and illustrative example of using the dual neural network to solve a QP problem. These above have substantiated the efficacy of the dual neural network for online QP problem solving.

References

1. Hopfield JJ, Tank DW (1985) Neural computation of decision in optimization problems. Biol Cybern 52:122–141
2. Kennedy MP, Chua LO (1988) Neural networks for nonlinear programming. IEEE Trans Circuits Syst 35(5):554–562
3. Zhang S, Constantinides AG (1992) Lagrange programming neural networks. IEEE Trans Circuits Syst 39(7):441–452
4. Gong D, Gen M, Yamazaki G, Xu W (1996) A modified ANN for convex programming with linear constraints. In: Proceedings of IEEE international conference on neural networks, pp 537–542
5. Xia Y (1996) A new neural network for solving linear and quadratic programming problems. IEEE Trans Neural Netw 7(6):1544–1547
6. Tao Q, Cao J, Xue M, Qiao H (2001) A high performance neural network for solving nonlinear programming problems with hybrid constraints. Phys Lett A 288(2):88–94
7. Zhang Y (2002) Analysis and design of recurrent neural networks and their applications to control and robotic systems. PhD dissertation, Chinese University of Hong Kong, Hong Kong
8. Zhang Y, Wang J, Xia Y (2003) A dual neural network for redundancy resolution of kinematically redundant manipulators subject to joint limits and joint velocity limits. IEEE Trans Neural Netw 14(3):658–667
9. Zhang Y, Wang J (2002) A dual neural network for convex quadratic programming subject to linear equality and inequality constraints. Phys Lett A 298:271–278
10. Zhang Y, Ge SS, Lee TH (2004) A unified quadratic programming based dynamical system approach to joint torque optimization of physically constrained redundant manipulators. IEEE Trans Syst Man Cybern, Part B, Cybern 34(5):2126–2132
11. Zhang Y, Ma S (2007) Minimum-energy redundancy resolution of robot manipulators unified by quadratic programming and its online solution. In: Proceedings of IEEE international conference on mechatronics and automation, pp 3232–3237

12. Zhang Y, Li Z, Tan H (2006) Inequality-based manipulator-obstacle avoidance using the LVI-based primal–dual neural network. In: Proceedings of IEEE international conference on robotics and biomimetics, pp 1459–1464
13. Zhang Y (2005) On the LVI-based primal–dual neural network for solving online linear and quadratic programming problems. In: Proceedings of American control conference, pp 1351–1356
14. Zhang Y (2006) Inverse-free computation for infinity-norm torque minimization of robot manipulators. Mechatronics 16:177–184
15. Zhang Y (2005) Revisit the analog computer and gradient-based neural system for matrix inversion. In: Proceedings of IEEE international symposium on intelligent control, pp 1411–1416
16. Zhang Y, Heng PA, Fu A (1999) Estimate of exponential convergence rate and exponential stability for neural networks. IEEE Trans Neural Netw 10(6):1487–1493
17. Zhang Y, Wang J, Xu Y (2002) A dual neural network for bi-criteria kinematic control of redundant manipulators. IEEE Trans Robot Autom 18(6):923–931
18. Zhang Y, Wang J (2002) A dual neural network for constrained joint torque optimization of kinematically redundant manipulators. IEEE Trans Syst Man Cybern, Part B, Cybern 32(5):654–662
19. Bazaraa MS, Sherali HD, Shetty CM (1993) Nonlinear programming—theory and algorithms. Wiley, New York
20. Wang J, Xia Y (1999) A dual neural network solving quadratic programming problems. In: Proceedings of international joint conference on neural networks, pp 588–593
21. Mangasarian OL (1979) Solution of symmetric linear complementarity problems by iterative methods. J Optim Theory Appl 22(2):465–485
22. Bertsekas DP (1989) Parallel and distributed computation: numerical methods. Prentice-Hall, New Jersey
23. Li W, Swetits J (1997) A new algorithm for solving strictly convex quadratic programs. SIAM J Optim 7(3):595–619
24. Zhang Y, Ge SS (2005) Design and analysis of a general recurrent neural network model for time-varying matrix inversion. IEEE Trans Neural Netw 16(6):1477–1490
25. Mead C (1989) Analog VLSI and neural systems. Addison-Wesley, Reading
26. Zhang Y (2007) Recurrent neural networks: design, analysis, applications to control and robotic systems. Lambert Academic, Germany
27. Zhang Y, Leithead WE, Leith DJ (2005) Time-series Gaussian process regression based on Toeplitz computation of $O(N^2)$ operations and $O(N)$-level storage. In: Proceedings of the 44th IEEE conference on decision and control, pp 3711–3716
28. Zhang Y, Jiang D, Wang J (2002) A recurrent neural network for solving Sylvester equation with time-varying coefficients. IEEE Trans Neural Netw 13(5):1053–1063
29. Zhang Y, Wang J (2002) Global exponential stability of recurrent neural networks for synthesizing linear feedback control systems via pole assignment. IEEE Trans Neural Netw 13(3):633–644
30. Zhang Y (2006) A set of nonlinear equations and inequalities arising in robotics and its online solution via a primal neural network. Neurocomputing 70(1–3):513–524
31. Pang JS (1987) A posterior error bounds for the linearly-constrained variational inequality problem. Math Oper Res 12(3):474–484
32. Pang JS, Yao YC (1995) On a generalization of a normal map and equation. SIAM J Control Optim 33:168–184
33. The MathWorks, Inc (2004) MATLAB 7. Natick, MA

Chapter 5
Primal–Dual Neural Networks

Abstract This chapter presents three primal–dual neural networks, i.e., a traditional primal–dual neural network (TPDNN), an LVI-based primal–dual neural network (LVI-PDNN) and a simplified LVI-PDNN. For demonstrating the wide applicability and effectiveness of such three neural networks, the traditional PDNN is used to solve online a linear program and its dual problem; the LVI-PDNN is used to solve the QP and LP problems; and the simplified LVI-PDNN is used to solve the strictly-convex QP problem subject to equality, inequality, and bound constraints.

5.1 Traditional Primal–Dual Neural Network

This section revisits the theory of a primal LP problem and its dual problem, which can be used to develop a kind of recurrent neural network for solving online LP problems as well as kinematic control of redundant manipulators [1] [e.g., the LP converted from (2.17)–(2.20)], i.e., a so-called traditional primal–dual neural network (TPDNN) initiated by Tang et al. However, due to the complexity of duality theory, that TPDNN needs to be improved so as to obtain the optimal solution(s) instead of feasible solutions. Computer simulation results substantiate the efficacy and correctness of the improved TPDNN model for online solution of LP problems.

5.1.1 Introduction

The LP problem is considered to be one of the basic problems widely encountered in a broad class of scientific disciplines and engineering applications [2–6]. In view of their fundamental roles, the LP problems have been investigated extensively for the past decades. Generally speaking, numerical algorithms have been employed popularly for solving such LP problems. However, it may not be efficient enough because of its serial-processing nature performed on digital computers [7]. Recently, due to the in-depth research in neural networks, numerous dynamic and analogue solvers based on recurrent neural networks have been investigated and developed [2–6, 8, 9]. The neural network approach is now regarded as a powerful alternative to

Y. Zhang, Z. Zhang, *Repetitive Motion Planning and Control of Redundant Robot Manipulators*, DOI 10.1007/978-3-642-37518-7_5,

online computation, owing to its parallel-distributed-processing nature and circuit-implementation convenience [8, 9].

Based on the duality theory [10, 11], a special kind of recurrent neural network (namely, traditional primal–dual neural network, TPDNN) was developed and exploited for kinematic control of redundant manipulators by Tang et al. [3, 4]. Section 5.1.2 presents an improved version of Tang et al.'s primal–dual neural network model, which solves for the optimal solution(s) of LP problems instead of feasible solutions. Two illustrative examples based on the MATLAB Simulink modeling platform [12] are presented in Sect. 5.1.3, which substantiate well the theoretical analysis and physical realizability.

Before ending this introductory subsection, it is worth mentioning here the main contributions of the section.

- By following the duality results presented in Sect. 4.1.2, an improved primal-dual neural network, different from the primal–dual neural network proposed by [3, 4], is exploited for solving online the linear-programming problems with optimal solution obtained rather than feasible solutions.
- Illustrative computer simulation results based on MATLAB Simulink platform substantiate further the efficacy and correctness of the improved TPDNN model for online LP solution. Moreover, the presented neural-network Simulink modeling approach appears to be an important and necessary step to FPGA and ASIC implementation of neural models, which could be done readily by using Simulink HDL Coder [12] as a future research direction.

5.1.2 TPDNN Model Description

Let us consider the following LP problem [3, 4] with $\hat{d} \in R^m$:

$$\begin{aligned} &\text{minimize } \hat{q}^{\mathrm{T}}\hat{x}, \\ &\text{subject to } J\hat{x} = \hat{d}, \\ &\qquad\qquad A\hat{x} \le \hat{b}, \end{aligned} \tag{5.1}$$

where $\hat{x} \in R^n$ is the primal decision variable vector to be solved for. By following the general formulation of primal and dual LP problems presented in Sect. 4.1.2, the corresponding dual problem of LP (5.1) can be written down readily as

$$\begin{aligned} &\text{maximize } \hat{d}^{\mathrm{T}}u - \hat{b}^{\mathrm{T}}v, \\ &\text{subject to } J^{\mathrm{T}}u - A^{\mathrm{T}}v = \hat{q}, \\ &\qquad\qquad u \text{ unrestricted}, \quad v \ge 0, \end{aligned} \tag{5.2}$$

where $u \in R^m$ and $v \in R^{\dim(\hat{b})}$ are the dual decision variable vectors defined respectively for equality and inequality constraints of LP (5.1).

By the duality theory of optimization [10, 11], if $\hat{x}^* \in R^n$ is an optimal solution to LP (5.1), then there exists a column vector $\hat{z}^* = [u^{*\mathrm{T}}, v^{*\mathrm{T}}]^{\mathrm{T}} \in R^{m+\dim(\hat{b})}$ that is an optimal solution to (5.2). Let the optimal-solution set $\Omega^* := \{\hat{y}^* = [\hat{x}^{*\mathrm{T}}, \hat{z}^{*\mathrm{T}}]^{\mathrm{T}} \mid \hat{x}^*$ solve (5.1), and $\hat{z}^*$ solve (5.2)$\}$. Then, a necessary and sufficient condition for solution $[\hat{x}^{\mathrm{T}}, u^{\mathrm{T}}, v^{\mathrm{T}}]^{\mathrm{T}} \in \Omega^*$ is

$$\begin{aligned} &\hat{q}^{\mathrm{T}}\hat{x} = \hat{d}^{\mathrm{T}}u - \hat{b}^{\mathrm{T}}v, \\ &J\hat{x} = \hat{d}, \qquad -A\hat{x} + \hat{b} \geq 0, \\ &J^{\mathrm{T}}u - A^{\mathrm{T}}v = \hat{q}, \quad v \geq 0. \end{aligned} \tag{5.3}$$

Thus, we can define the following energy function for TPDNN design:

$$\begin{aligned} \varepsilon(\hat{x}, u, v) = {} & \frac{1}{2}\big(\hat{q}^{\mathrm{T}}\hat{x} - \hat{d}^{\mathrm{T}}u + \hat{b}^{\mathrm{T}}v\big)^2 + \frac{1}{2}\|J\hat{x} - \hat{d}\|_2^2 \\ & + \frac{1}{4}(-A\hat{x} + \hat{b})^{\mathrm{T}}\big((-A\hat{x} + \hat{b}) - |-A\hat{x} + \hat{b}|\big) \\ & + \frac{1}{2}\|J^{\mathrm{T}}u - A^{\mathrm{T}}v - \hat{q}\|_2^2 \\ & + \frac{1}{4}v^{\mathrm{T}}\big(v - |v|\big) \geq 0. \end{aligned} \tag{5.4}$$

It is worth pointing out that the energy function (5.4) is convex and continuously differentiable [2–4]. In addition, the minimum value $\varepsilon(\hat{x}^*, u^*, v^*) = 0$ is achieved if and only if $\hat{x}^*$ and $\hat{z}^* = [u^{*\mathrm{T}}, v^{*\mathrm{T}}]^{\mathrm{T}}$ are the optimal solutions of primal LP (5.1) and its dual problem (5.2), respectively. Moreover, based on such an energy function (5.4), the dynamic equations of the traditional primal–dual neural network model [which solves online (5.1) and (5.2) simultaneously] can be designed as follows by using the negative-gradient descent direction:

$$\begin{aligned} \frac{\mathrm{d}\hat{x}}{\mathrm{d}t} &= -\bar{\gamma}\frac{\partial\varepsilon(\hat{x}, u, v)}{\partial\hat{x}}, \\ \frac{\mathrm{d}u}{\mathrm{d}t} &= -\bar{\gamma}\frac{\partial\varepsilon(\hat{x}, u, v)}{\partial u}, \\ \frac{\mathrm{d}v}{\mathrm{d}t} &= -\bar{\gamma}\frac{\partial\varepsilon(\hat{x}, u, v)}{\partial v}, \end{aligned} \tag{5.5}$$

where the design parameter $\bar{\gamma} > 0$, as before, is used to scale the convergence rate of the recurrent neural network and should be set as large as the hardware would permit.

Define $\psi(\tilde{x}_i) = \min\{0, \tilde{x}_i\}$ for any element $\tilde{x}_i$ of column vector $\tilde{x} \in R^n$, and $\Psi(\tilde{x}) := [\psi(\tilde{x}_1), \psi(\tilde{x}_2), \ldots, \psi(\tilde{x}_n)]^{\mathrm{T}}$. Then, we have $\tilde{x} - |\tilde{x}| = 2\Psi(\tilde{x})$. Therefore,

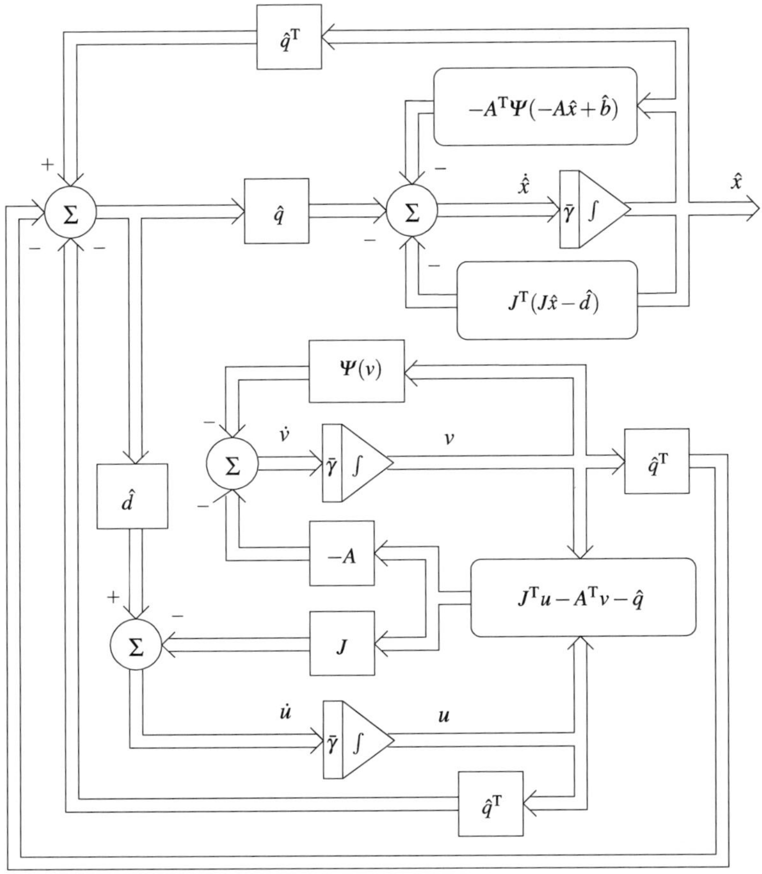

Fig. 5.1 Block diagram of TPDNN model (5.6) for solving online LP (5.1)

the neural dynamics (5.5) can be expanded as

$$
\begin{aligned}
\frac{\mathrm{d}\hat{x}}{\mathrm{d}t} &= -\bar{\gamma}\big(\hat{q}\big(\hat{q}^{\mathrm{T}}\hat{x}-\hat{d}^{\mathrm{T}}u+\hat{b}^{\mathrm{T}}v\big)+J^{\mathrm{T}}(J\hat{x}-\hat{d})-A^{\mathrm{T}}\Psi(-A\hat{x}+\hat{b})\big),\\
\frac{\mathrm{d}u}{\mathrm{d}t} &= -\bar{\gamma}\big(-\hat{d}\big(\hat{q}^{\mathrm{T}}\hat{x}-\hat{d}^{\mathrm{T}}u+\hat{b}^{\mathrm{T}}v\big)+J\big(J^{\mathrm{T}}u-A^{\mathrm{T}}v-\hat{q}\big)\big),\\
\frac{\mathrm{d}v}{\mathrm{d}t} &= -\bar{\gamma}\big(\hat{b}\big(\hat{q}^{\mathrm{T}}\hat{x}-\hat{d}^{\mathrm{T}}u+\hat{b}^{\mathrm{T}}v\big)-A\big(J^{\mathrm{T}}u-A^{\mathrm{T}}v-\hat{q}\big)+\Psi(v)\big).
\end{aligned}
\tag{5.6}
$$

In addition, the block diagram of traditional primal–dual neural network (5.6) [also termed TPDNN (5.6)] is depicted in Fig. 5.1. It shows that such a neural network uses only simple hardware (i.e., adders, multipliers, integrators, and the $\min\{0,\tilde{x}_i\}$ circuits). Furthermore, by assuming the existence of at least one optimal solution to LP (5.1) and as generalized from [2–4, 10], for the proposed traditional primal–dual

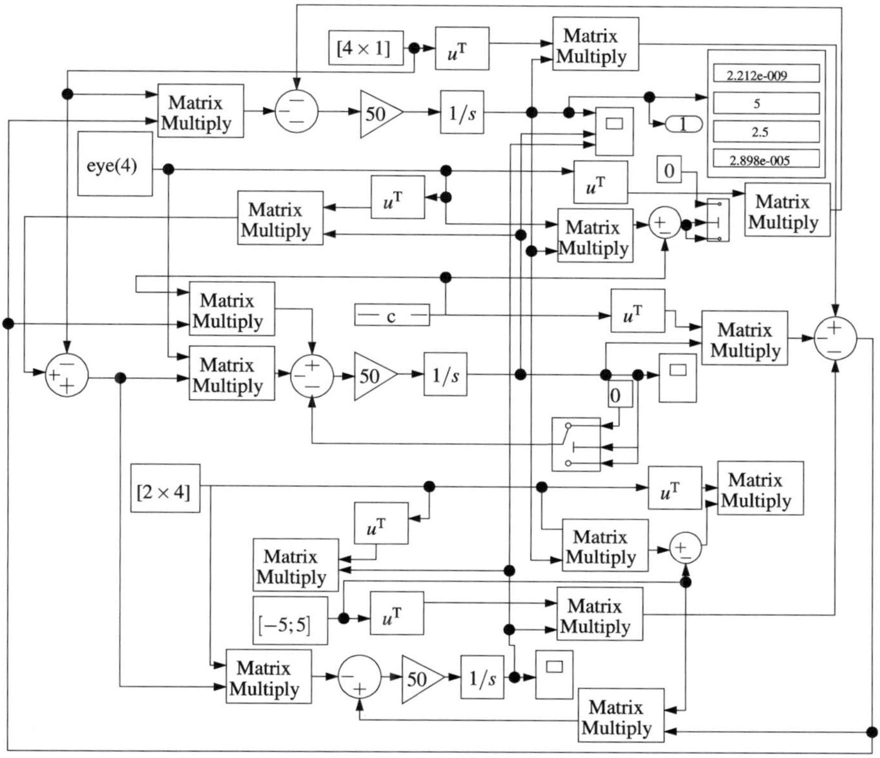

Fig. 5.2 Overall Simulink model of traditional primal–dual neural network (5.6)

neural network (5.6), the following theoretical results are presented about its global convergence and stability.

Proposition 5.1 *The optimal solutions to LP* (5.1) *and its dual LP problem* (5.2) *are respectively* $\hat{x}^*$ *and* $\hat{z}^* = [u^{*\mathrm{T}}, v^{*\mathrm{T}}]^{\mathrm{T}}$ *if and only if the energy function* (5.4) *satisfies* $\varepsilon(\hat{x}^*, u^*, v^*) = 0$.

Proposition 5.2 *The traditional primal–dual neural network* (5.6) *is globally stable, and its state* $\hat{y} := [\hat{x}^{\mathrm{T}}, u^{\mathrm{T}}, v^{\mathrm{T}}]^{\mathrm{T}}$ *can always converge to an optimal solution* $\hat{y}^* = [\hat{x}^{*\mathrm{T}}, u^{*\mathrm{T}}, v^{*\mathrm{T}}]^{\mathrm{T}}$ *of the linear program* (5.1) *and its dual linear program* (5.2).

5.1.3 Computer Simulation Study

To demonstrate the behavior and efficacy of the traditional primal–dual neural-network model (5.6), we investigate two illustrative examples based on MATLAB Simulink modeling technique [12], with the overall TPDNN model shown in Fig. 5.2 (note that the values correspond to the following example).

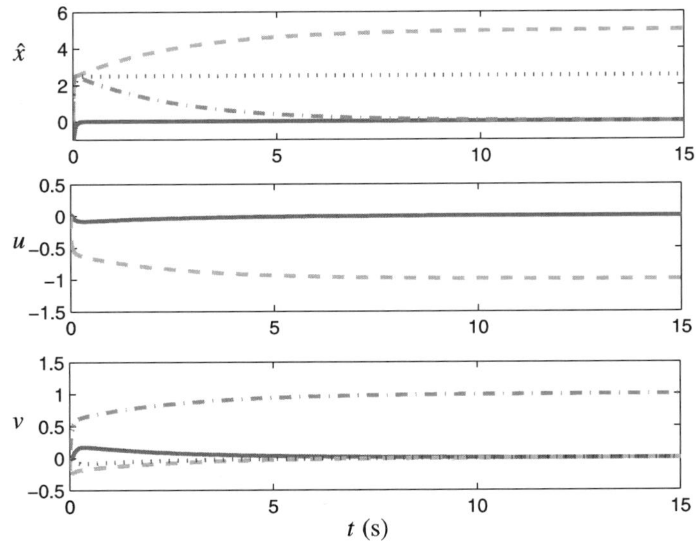

Fig. 5.3 Online solution of TPDNN (5.6) starting from zero initial states with $\bar{\gamma} = 50$

Example 5.1 Consider the following LP problem:

$$\begin{aligned} &\text{minimize } -\hat{x}_2 + 6, \\ &\text{subject to } 2\hat{x}_1 - 2\hat{x}_3 = -5, \\ &\qquad\qquad \hat{x}_2 + \hat{x}_4 = 5, \\ &\qquad\qquad \hat{x}_i \geq 0, \quad i = 1, 2, 3, 4, \end{aligned}$$

which could be rewritten in the compact matrix form (5.1), and we thus have the coefficients $\hat{q} = [0, -1, 0, 0]^{\mathrm{T}}$, $J = [2, 0, -2, 0; 0, 1, 0, 1] \in R^{2\times 4}$ (in MATLAB notation), $\hat{d} = [-5, 5]^{\mathrm{T}}$, $A = -I \in R^{4\times 4}$ with I denoting the identity matrix, and $\hat{b} = 0 \in R^4$.

It is worth mentioning here that the optimal solution (or to say, minimizer) to the above primal LP problem is multiple and unbounded, and that, for comparison purposes, its set could be given analytically as $\Omega^*_{\hat{x}} := \{\hat{x}^* := [\hat{x}^*_1, 5, \hat{x}^*_3, 0] \mid 2\hat{x}^*_1 - 2\hat{x}^*_3 = -5, \hat{x}^*_1 > 0, \hat{x}^*_3 > 0\}$. By solving this LP problem via traditional primal–dual neural network (5.6) with design parameter $\bar{\gamma} = 50$, computer simulation results are summarized below.

- Starting from zero initial states $\hat{x}(0) = 0 \in R^4$, $u(0) = 0 \in R^2$, and $v(0) = 0 \in R^4$, the neural state $\hat{y} = [\hat{x}^{\mathrm{T}}, u^{\mathrm{T}}, v^{\mathrm{T}}]^{\mathrm{T}}$ of TPDNN (5.6) can converge to the optimal solution $\hat{y}^* = [\hat{x}^{*\mathrm{T}}, u^{*\mathrm{T}}, v^{*\mathrm{T}}]^{\mathrm{T}}$, where $\hat{x}^* = [0, 5, 2.5, 0]^{\mathrm{T}}$, $u^* = [0, -1]^{\mathrm{T}}$, and $v^* = [0, 0, 0, 1]^{\mathrm{T}}$, as illustrated in Fig. 5.3.
- Starting from different initial states such as $\hat{x}(0) = [3, 8, 10, 0.5]^{\mathrm{T}}$, $u(0) = 0 \in R^2$, and $v(0) = 0 \in R^4$, the neural state $\hat{y} = [\hat{x}^{\mathrm{T}}, u^{\mathrm{T}}, v^{\mathrm{T}}]^{\mathrm{T}}$ of TPDNN (5.6) converges to an optimal solution $\hat{y}^* = [\hat{x}^{*\mathrm{T}}, u^{*\mathrm{T}}, v^{*\mathrm{T}}]^{\mathrm{T}}$ as well, where

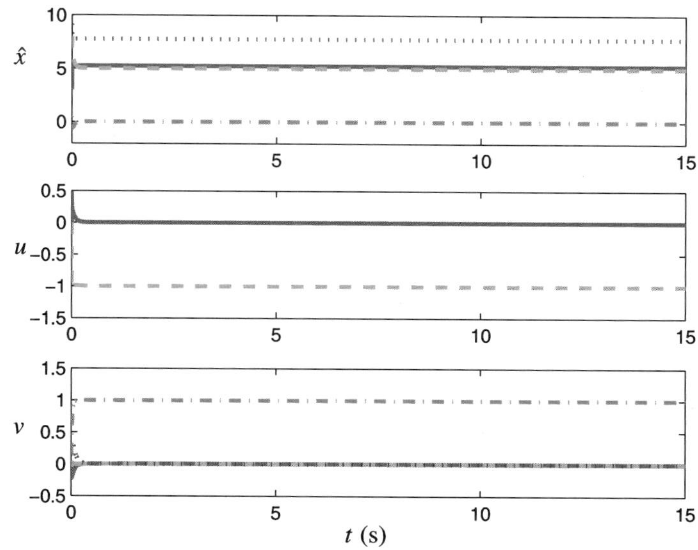

Fig. 5.4 Online solution of TPDNN (5.6) starting from different initial states with $\bar{\gamma} = 50$

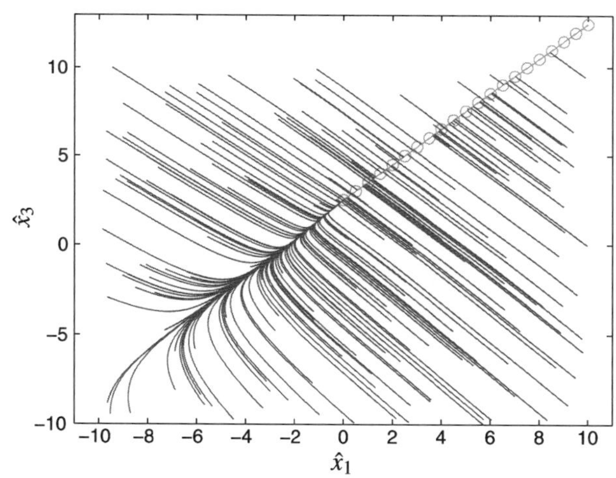

Fig. 5.5 Trajectories of neural states $\hat{x}_1$ and $\hat{x}_3$ of TPDNN model (5.6) with $\bar{\gamma} = 50$, starting from 200 randomly generated initial states and converging to the optimal-solution set $\Omega_{\hat{x}}^*$ denoted by O

$\hat{x}^* = [5.25, 5, 7.75, 0]^{\mathrm{T}}$, $u^* = [0, -1]^{\mathrm{T}}$, and $v^* = [0, 0, 0, 1]^{\mathrm{T}}$. This is shown in Fig. 5.4.

- In addition, the planar trajectories of neural states $\hat{x}_1(t)$ and $\hat{x}_3(t)$ of the traditional primal–dual neural network (5.6) that start from randomly generated initial states within $[-10, 10]$ are shown in Fig. 5.5. Evidently, the trajectories all converge to the optimal-solution set $\Omega_{\hat{x}}^*$ denoted by O in the figure.

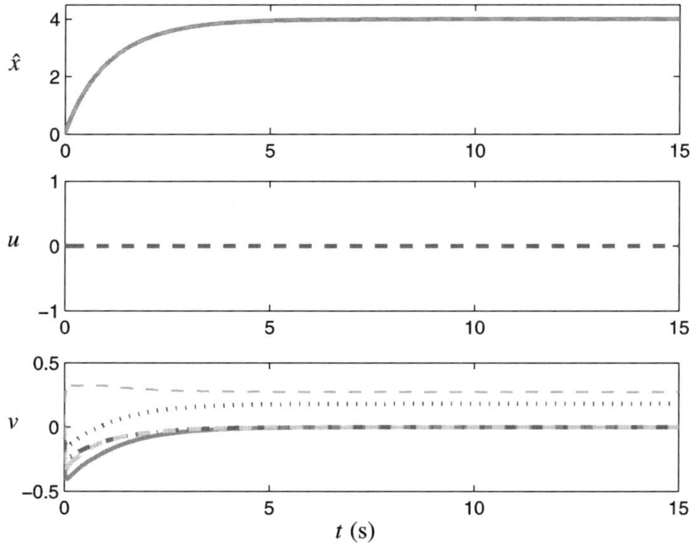

Fig. 5.6 Online solution of TPDNN model (5.6) starting from zero initial states with $\bar{\gamma} = 50$

Example 5.2 Consider the online solution of the following LP problem by using the above-proposed traditional primal–dual neural network (5.6):

$$\begin{aligned} &\text{minimize } -\hat{x}_1 - \hat{x}_2, \\ &\text{subject to } -2\hat{x}_1 + \hat{x}_2 \leq 3, \\ &\qquad \hat{x}_1 + 3\hat{x}_2 \leq 16, \\ &\qquad 4\hat{x}_1 + \hat{x}_2 \leq 20, \\ &\qquad \hat{x}_1, \hat{x}_2 \geq 0, \end{aligned}$$

which can be rewritten in the matrix form (5.1) with the coefficients $\hat{q} = [-1, -1]^{\mathrm{T}}$, $J = [0, 0]$, $\hat{d} = 0$, $A = [-2, 1; 1, 3; 4, 1; -1, 0; 0, -1] \in R^{5\times 2}$, and $\hat{b} = [3, 16, 20, 0, 0]^{\mathrm{T}}$. Then the transient behavior and performance of TPDNN (5.6) when solving this LP problem can be simulated readily and summarized as follows.

- As illustrated in Fig. 5.6, starting from zero initial states $\hat{x}(0) = 0 \in R^2$, $u(0) = 0 \in R$, and $v(0) = 0 \in R^5$, the state vector $\hat{x}(t) \in R^2$ of the traditional primal–dual neural network (5.6) converges to the unique optimal solution $\hat{x}^* = [4, 4]^{\mathrm{T}}$ of the above LP problem.
- As illustrated in Fig. 5.7, the neural state $\hat{x}(t)$ of the traditional primal–dual neural network (5.6), starting from randomly generated initial states within $[-10, 10]$, always converges to the optimal solution $\hat{x}^* = [4, 4]^{\mathrm{T}}$ of the above LP problem (denoted by O).

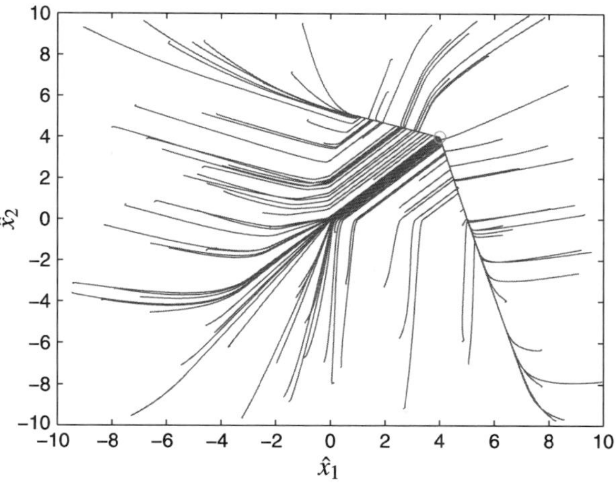

Fig. 5.7 Trajectories of $\hat{x}_1$ and $\hat{x}_2$ of the TPDNN model (5.6) with $\bar{\gamma} = 50$, starting from 100 randomly generated initial states and converging to equilibrium point $\hat{x}^* = [4, 4]^{\mathrm{T}}$ denoted by O, which is also the theoretically optimal solution to LP (5.1) as in Example 5.2

In summary, the above illustrative examples can substantiate further the global convergence and efficacy of the improved traditional primal–dual neural network (5.6) on linear-program solving.

5.2 LVI-Based Primal–Dual Neural Networks

Motivated by real-time solution to robotic problems, researchers have to consider the general unified formulation of linear and quadratic programs subject to equality, inequality, and bound constraints simultaneously. A primal–dual neural network is presented in this section for the online solution based on the linear variational inequality (LVI) [13]. The neural network is of simple piecewise-linear dynamics, globally convergent to optimal solutions, and is able to handle linear and quadratic problems in the same manner. Other robotics-related properties of the LVI-based primal–dual network are also investigated, such as the convergence starting within feasible regions and the case of no solutions.

5.2.1 General Problem Formulation

In view of its fundamental role arising in numerous fields of science and engineering, the problem of solving linear and quadratic programs has been investigated extensively for the past decades. In the literature, researches are usually solving LP and QP problems separately. In addition, they handle optimization problems only subject to one or two kinds of constraints [14]. Motivated by engineering applications of linear/quadratic program in robotics [15–20], however, the following general problem formulation is preferred, the same as in the previous chapters:

$$\text{minimize } \hat{x}^{\mathrm{T}}Q\hat{x}/2+\hat{q}^{\mathrm{T}}\hat{x}, \tag{5.7}$$

$$\text{subject to } J\hat{x}=\hat{d}, \tag{5.8}$$

$$A\hat{x}\le\hat{b}, \tag{5.9}$$

$$\xi^{-}\le\hat{x}\le\xi^{+}, \tag{5.10}$$

where Q is assumed only positive semi-definite in this chapter such that QP and LP are both to be handled in this formulation.

Before proceeding, we briefly review the following existing neural networks. It is known that the early neural model like [21] contains finite penalty parameters and generates approximate solutions only. The Lagrangian neural network may have premature defect when applied to inequality-constrained QP problems [22]. To always obtain optimal/exact solutions, traditional primal–dual neural networks were proposed based on the Karush–Kuhn–Tacker conditions and projection operator [3]; please see also the one in the previous subsection. However, due to minimizing the duality gap by gradient descent methods, the dynamic equations of such primal–dual networks are usually complicated, even containing high-order nonlinear terms [14]. To reduce implementation and computation complexities, a dual neural network (e.g., the one in the previous chapter) has been finally developed for handling general QP (5.7)–(5.10) with simple piecewise linearity and global convergence to optimal solutions [18]. The disadvantage of dual networks, however, is that they require the inversion of coefficient matrix Q and thus are only able to handle strictly convex quadratic programs, preferably with an easy-to-invert matrix Q [14].

As the research evolves spirally, a primal–dual neural network model has recently been "discovered" with simple piecewise linear dynamics, global convergence to optimal solutions, and capability of handling QP and LP problems in the same/unified manner [15, 20]. Because the primal–dual neural network model is designed based on LVI (linear variational inequality), it is termed the LVI-based primal–dual neural network. This section presents the relatively whole picture of the LVI-based primal–dual network for solving QP/LP (5.7)–(5.10), including rigorous derivation, convergence properties, and solution characteristics.

5.2.2 Reformulation and LVI-PDNN Solution

For developing the LVI-based primal–dual neural network to solve online (5.7)–(5.10), we can firstly convert QP/LP (5.7)–(5.10) to the LVI (note that the detailed derivation can be seen from Theorem 5.2 in Sect. 5.2.3). That is, to find a vector $\hat{y}^{*}\in\Omega$ such that, for all $\hat{y}\in\Omega:=\{\hat{y}\mid\varsigma^{-}\le\hat{y}\le\varsigma^{+}\}\subset R^{n+m+\dim(\hat{b})}$,

$$(\hat{y}-\hat{y}^{*})^{\mathrm{T}}(H\hat{y}^{*}+\hat{p})\ge 0, \tag{5.11}$$

where the primal–dual decision vector $\hat{y}$ and its lower/upper bounds are defined respectively as

$$\hat{y}=\begin{bmatrix}\hat{x}\\u\\v\end{bmatrix},\qquad \varsigma^{-}=\begin{bmatrix}\xi^{-}\\-\varpi\\0\end{bmatrix},\qquad \varsigma^{+}=\begin{bmatrix}\xi^{+}\\\varpi\\\varpi\end{bmatrix},\tag{5.12}$$

with ϖ representing a vector of appropriate dimensions whose elements are sufficiently large positive constants. The coefficients in (5.11) are defined as

$$H=\begin{bmatrix}Q & -J^{\mathrm{T}} & A^{\mathrm{T}}\\J & 0 & 0\\-A & 0 & 0\end{bmatrix},\qquad \hat{p}=\begin{bmatrix}\hat{q}\\-\hat{d}\\\hat{b}\end{bmatrix}.\tag{5.13}$$

Moreover, LVI (5.11) is equivalent to the following piecewise linear equation (PLE) (note that the detailed derivation can be seen from Theorem 5.3 in Sect. 5.2.3):

$$\mathscr{P}_{\Omega}\big(\hat{y}-(H\hat{y}+\hat{p})\big)-\hat{y}=0,\tag{5.14}$$

where a piecewise-linear projection operator $\mathscr{P}_{\Omega}(\cdot):R^{n+m+\dim(\hat{b})}\rightarrow\Omega=\{\hat{y}\mid\varsigma^{-}\le\hat{y}\le\varsigma^{+}\}$ is extensively used and described as $\mathscr{P}_{\Omega}(\hat{y}):=[[\mathscr{P}_{\Omega}(\hat{y})]_1,[\mathscr{P}_{\Omega}(\hat{y})]_2,\ldots,[\mathscr{P}_{\Omega}(\hat{y})]_{n+m+\dim(\hat{b})}]^{\mathrm{T}}$ with the ith processing element defined as

$$\big[\mathscr{P}_{\Omega}(\hat{y})\big]_i=\begin{cases}\varsigma_i^{-} & \text{if } \hat{y}_i<\varsigma_i^{-},\\ \hat{y}_i & \text{if } \varsigma_i^{-}\le\hat{y}_i\le\varsigma_i^{+},\\ \varsigma_i^{+} & \text{if } \hat{y}_i>\varsigma_i^{+}.\end{cases}\quad i=1,2,\ldots,n+m+\dim(\hat{b}),$$

For graphical interpretation of the ith processing element $[\mathscr{P}_{\Omega}(\hat{y})]_i$, see Fig. 5.8, which is actually very similar to Fig. 4.1.

It follows naturally from our neural-network design experience [14–20, 23–25] that the LVI-based primal–dual neural network solver can have the following dynamic equation:

$$\dot{\hat{y}}=\bar{\gamma}\big(I+H^{\mathrm{T}}\big)\big(\mathscr{P}_{\Omega}\big(\hat{y}-(H\hat{y}+\hat{p})\big)-\hat{y}\big),\tag{5.15}$$

where $\bar{\gamma}>0$ is the design parameter used to scale the network convergence.

The block diagram realizing (5.15) is presented in Fig. 5.9, wherein a piecewise-linear activation function $\mathscr{P}_{\Omega}(\cdot)$ can be implemented by using operational amplifiers known as limiters [14]. As bound constraint (5.10) is neatly cast into projection set Ω, the size of the LVI-based primal–dual neural network (5.15) is only the dimension sum of the equality constraint (5.8), inequality constraint (5.9), and primal decision vector $\hat{x}$, which is much smaller than those in [3, 22]. In addition, the LVI-based primal–dual network does not involve matrix inversion, matrix–matrix multiplication, or high-order nonlinear computation, thus reducing the implementation and computation complexities, as compared to other recurrent neural models

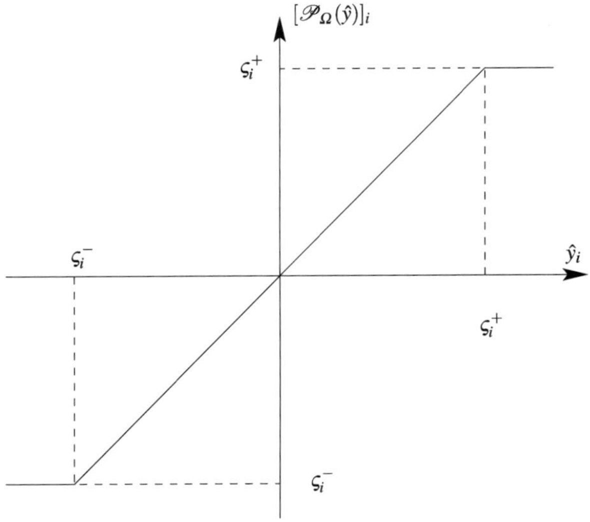

Fig. 5.8 The ith processing element of piecewise-linear operator $\mathscr{P}_\Omega(\cdot)$

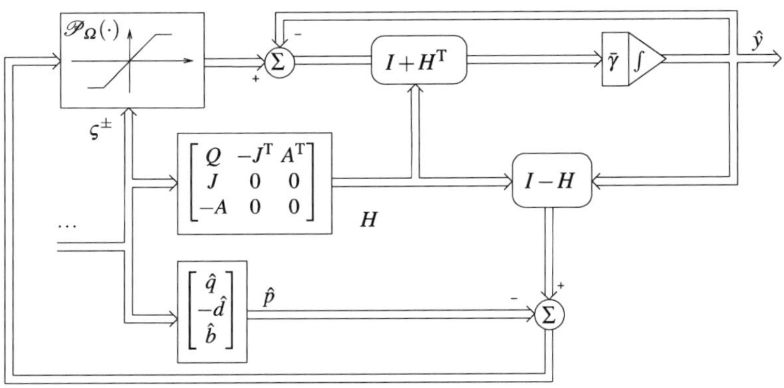

Fig. 5.9 Block diagram of LVI-based PDNN for solving (5.7)–(5.10)

[3, 14]. Furthermore, we have the following important results on the LVI-PDNN convergence [13, 15].

Theorem 5.1 *Starting from any initial state, the state vector* $\hat{y}(t)$ *of the primal-dual neural network* (5.15) *converges to an equilibrium point* $\hat{y}^*$, *of which the first n elements constitute the optimal solution* $\hat{x}^*$ *to the QP problem* (5.7)–(5.10). *Moreover, the exponential convergence can be achieved, provided that there exists a constant* $\varrho > 0$ *such that*

$$\big\|\hat{y} - \mathscr{P}_\Omega\big(\hat{y} - (H\hat{y} + \hat{p})\big)\big\|_2^2 \geq \varrho\big\|\hat{y} - \hat{y}^*\big\|_2^2.$$

Proof The proof can be generalized from [17, 18, 24]. To show the convergence, the finally useful inequality (5.19) needs to be obtained through (5.16) and (5.17).

First, it follows from the projection inequality $(\mathscr{P}_\Omega(\omega)-\varpi)^{\mathrm{T}}(\omega-\mathscr{P}_\Omega(\omega))\geq 0$ for all $\omega\in R^{n+m+dim(\hat{b})}$ and $\varpi\in\Omega$ [17, 24] that

$$\big(\mathscr{P}_\Omega\big(\hat{y}-(H\hat{y}+\hat{p})\big)-\hat{y}^*\big)^{\mathrm{T}}\big(\hat{y}-(H\hat{y}+\hat{p})-\mathscr{P}_\Omega\big(\hat{y}-(H\hat{y}+\hat{p})\big)\big)\geq 0,$$

or written as follows for consistency:

$$\big(\hat{y}^*-\mathscr{P}_\Omega\big(\hat{y}-(H\hat{y}+\hat{p})\big)\big)^{\mathrm{T}}\big(H\hat{y}+\hat{p}-\hat{y}+\mathscr{P}_\Omega\big(\hat{y}-(H\hat{y}+\hat{p})\big)\big)\geq 0. \quad (5.16)$$

Second, it follows from the projection-equation reformulation of the linear variational inequality, (5.14), that

$$\big(\hat{y}^*-\mathscr{P}_\Omega\big(\hat{y}-(H\hat{y}+\hat{p})\big)\big)^{\mathrm{T}}\big(-H\hat{y}^*+\hat{p}\big)\geq 0. \quad (5.17)$$

Summing up (5.16) and (5.17) yields

$$\big(\hat{y}^*-\mathscr{P}_\Omega\big(\hat{y}-(H\hat{y}+\hat{p})\big)\big)^{\mathrm{T}}\big(H\big(\hat{y}-\hat{y}^*\big)-\hat{y}+\mathscr{P}_\Omega\big(\hat{y}-(H\hat{y}+\hat{p})\big)\big)\geq 0$$

and

$$\begin{aligned}&\big(\hat{y}^*-\hat{y}+\hat{y}-\mathscr{P}_\Omega\big(\hat{y}-(H\hat{y}+\hat{p})\big)\big)^{\mathrm{T}}\\&\quad\times\big(H\big(\hat{y}-\hat{y}^*\big)-\hat{y}+\mathscr{P}_\Omega\big(\hat{y}-(H\hat{y}+\hat{p})\big)\big)\geq 0. \end{aligned}\quad (5.18)$$

Then, extending (5.18) gives

$$\begin{aligned}&\big(\hat{y}-\hat{y}^*\big)^{\mathrm{T}}\big(I+H^{\mathrm{T}}\big)\big(\hat{y}-\mathscr{P}_\Omega\big(\hat{y}-(H\hat{y}+\hat{p})\big)\big)\\&\geq\big\|\hat{y}-\mathscr{P}_\Omega\big(\hat{y}-(H\hat{y}+\hat{p})\big)\big\|_2^2+\big(\hat{y}-\hat{y}^*\big)^{\mathrm{T}}H\big(\hat{y}-\hat{y}^*\big).\end{aligned}$$

Considering that H is positive semidefinite (not necessarily symmetric) [24, 26] in terms of

$$\hat{y}^{\mathrm{T}}H\hat{y}=\hat{y}^{\mathrm{T}}\frac{H+H^{\mathrm{T}}}{2}\hat{y}=\hat{y}^{\mathrm{T}}\begin{bmatrix}Q&0\\0&0\end{bmatrix}\hat{y}\geq 0,$$

we have

$$\begin{aligned}&\big(\hat{y}-\hat{y}^*\big)^{\mathrm{T}}\big(I+H^{\mathrm{T}}\big)\big(\hat{y}-\mathscr{P}_\Omega\big(\hat{y}-(H\hat{y}+\hat{p})\big)\big)\\&\quad\geq\big\|\hat{y}-\mathscr{P}_\Omega\big(\hat{y}-(H\hat{y}+\hat{p})\big)\big\|_2^2+\big\|\hat{y}-\hat{y}^*\big\|_{\mathrm{H}}^2\geq 0.\end{aligned}\quad (5.19)$$

Define the Lyapunov function $V(\hat{y})=\|\hat{y}-\hat{y}^*\|_2^2\geq 0$. Its time derivative along the primal–dual neural network trajectory (5.15) is

$$\begin{aligned}\frac{\mathrm{d}V(\hat{y})}{\mathrm{d}t}&=\left(\frac{\partial V(\hat{y})}{\partial\hat{y}}\right)^{\mathrm{T}}\frac{\mathrm{d}\hat{y}}{\mathrm{d}t}\\&=\big(\hat{y}-\hat{y}^*\big)^{\mathrm{T}}\bar{\gamma}\big(I+H^{\mathrm{T}}\big)\big(\mathscr{P}_\Omega\big(\hat{y}-(H\hat{y}+\hat{p})\big)-\hat{y}\big)\end{aligned}$$

$$
\begin{aligned}
&= -\bar{\gamma}(\hat{y} - \hat{y}^*)^{\mathrm{T}}(I + H^{\mathrm{T}})(\hat{y} - \mathscr{P}_\Omega(\hat{y} - (H\hat{y} + \hat{p}))) \\
&\leq -\bar{\gamma}\|\hat{y} - \mathscr{P}_\Omega(\hat{y} - (H\hat{y} + \hat{p}))\|_2^2 - \bar{\gamma}\|\hat{y} - \hat{y}^*\|_{\mathrm{H}}^2 \\
&\leq 0.
\end{aligned} \tag{5.20}
$$

By Lyapunov theory, the network state $\hat{y}(t)$ is stable and globally convergent to an equilibrium $\hat{y}^*$ in view of $\dot{V} = 0$ for $\dot{\hat{y}} = 0$ and $\hat{y} = \hat{y}^*$. It follows from the conversion of QP/LP, LVI, and PLE (with detailed derivations in Theorems 5.2 and 5.3 in Sect. 5.2.3) that $\hat{y}^*$ is the solution to the linear variational inequality problem (5.11) and the first n elements of $\hat{y}^*$ constitute the optimal solution $\hat{x}^*$ to quadratic program (5.7)–(5.10).

As for the exponential convergence, let us review $V(\hat{y})$ and $\dot{V}(\hat{y})$. From (5.20) and the extra condition called the exponential-convergence condition (i.e., if there exists $\varrho > 0$ such that $\|\hat{y} - \mathscr{P}_\Omega(\hat{y} - (H\hat{y} + \hat{p}))\|_2^2 \geq \varrho\|\hat{y} - \hat{y}^*\|_2^2$) we have

$$
\begin{aligned}
\frac{\mathrm{d}V(\hat{y})}{\mathrm{d}t} &\leq -\bar{\gamma}\|\hat{y} - \mathscr{P}_\Omega(\hat{y} - (H\hat{y} + \hat{p}))\|_2^2 - \bar{\gamma}\|\hat{y} - \hat{y}^*\|_{\mathrm{H}}^2 \\
&\leq -\bar{\gamma}\varrho\|\hat{y} - \hat{y}^*\|_2^2 - \bar{\gamma}\|\hat{y} - \hat{y}^*\|_{\mathrm{H}}^2 \\
&= -\bar{\gamma}(\hat{y} - \hat{y}^*)^{\mathrm{T}}(\varrho I + H)(\hat{y} - \hat{y}^*) \\
&\leq -\lambda V(\hat{y}),
\end{aligned}
$$

where $\lambda = \varrho\bar{\gamma} > 0$ is the convergence rate. Note that the existence of the exponential-convergence condition can be justified in practice by considering the equivalence of $\hat{y} - \mathscr{P}_\Omega(\hat{y} - (H\hat{y} + \hat{p})) = 0$ and $\hat{y} = \hat{y}^*$ and that it is analyzed in [14, 17, 19] and the references therein. Besides, a similar exponential-convergence condition is exploited and investigated for dual neural network in the proof of Theorem 4.7 and Remarks 4.8 through 4.10 of Sect. 4.3.2 of Chap. 4. Thus, we have $V(\hat{y}) = O(\exp(-\lambda(t - t_0)))$ for all $t \geq t_0$, and hence $\|\hat{y} - \hat{y}^*\|_2 = O(\exp(-\lambda(t - t_0)/2))$, $t \geq t_0$. The exponential convergence property of this LVI-PDNN is thus proved. □

5.2.3 *Additional Theoretical Results*

For better usability and clarity, the general problem formulation (5.7)–(5.10) and its LVI-based primal-dual neural network solver (5.15) have been presented in the previous subsection with theoretical-analysis results, while the additional detailed derivation and theoretical analysis about the conversion of QP/LP, LVI, and PLE [13, 15, 27] are separated from them and presented now. Note that, as a basis of the theoretical analysis, the existence of at least one optimal solution $\hat{x}^*$ to the QP problem (5.7)–(5.10) is assumed throughout the section and also the book unless stated otherwise.

Theorem 5.2 *The QP/LP problem* (5.7)–(5.10) *can be reformulated as the LVI problem* (5.11).

Proof It follows from Theorem 4.1 that the Lagrangian dual problem of (5.7)–(5.10) can be derived as

$$\text{minimize} \quad -\frac{1}{2}\hat{x}^{\mathrm{T}}Q\hat{x}+\hat{d}^{\mathrm{T}}u-\hat{b}^{\mathrm{T}}v+\xi^{-\mathrm{T}}\underline{\nu}-\xi^{+\mathrm{T}}\overline{\nu}, \tag{5.21}$$

$$\text{subject to} \quad Q\hat{x}+\hat{q}-J^{\mathrm{T}}u+A^{\mathrm{T}}v-\underline{\nu}+\overline{\nu}=0, \tag{5.22}$$

$$\text{with } u \text{ unrestricted}, \quad v\geq 0,\ \underline{\nu}\geq 0,\ \overline{\nu}\geq 0. \tag{5.23}$$

where u, v, $\underline{\nu}$, and $\overline{\nu}\in R^n$ are the corresponding dual decision vectors. Then, a necessary and sufficient condition for optimum $(\hat{x}^*,u^*,v^*,\underline{\nu}^*,\overline{\nu}^*)$ of the primal QP problem (5.7)–(5.10) and its dual QP problem (5.21)–(5.23) is [13, 15]:
Primal feasibility:

$$J\hat{x}^*-\hat{d}=0, \tag{5.24}$$

$$\begin{aligned} &-A\hat{x}^*+\hat{b}\geq 0,\\ &\xi^-\leq\hat{x}^*\leq\xi^+; \end{aligned} \tag{5.25}$$

Dual feasibility:

$$\begin{aligned} &Q\hat{x}^*+\hat{q}-J^{\mathrm{T}}u^*+A^{\mathrm{T}}v^*-\underline{\nu}^*+\overline{\nu}^*=0,\\ &u^* \text{ unrestricted}, \quad v^*\geq 0,\ \underline{\nu}^*\geq 0,\ \overline{\nu}^*\geq 0; \end{aligned} \tag{5.26}$$

Complementarity:

$$v^{*\mathrm{T}}\big(-A\hat{x}^*+\hat{b}\big)=0, \tag{5.27}$$

$$\underline{\nu}^{*\mathrm{T}}\big(-\hat{x}^*+\xi^-\big)=0, \tag{5.28}$$

$$\overline{\nu}^{*\mathrm{T}}\big(-\xi^++\hat{x}^*\big)=0. \tag{5.29}$$

To simplify the above necessary and sufficient formulation, we further study dual variable vectors $\underline{\nu}^*$ and $\overline{\nu}^*$ in (5.26), (5.28), and (5.29), which correspond to the bound constraint (5.10).

By defining $\nu^*=\underline{\nu}^*-\overline{\nu}^*$, the dual feasibility constraint (5.26) becomes

$$\big[Q\hat{x}^*+\hat{q}-J^{\mathrm{T}}u^*+A^{\mathrm{T}}v^*\big]_i=\nu_i^*\begin{cases}\leq 0 & \text{if } \hat{x}_i^*=\xi_i^+,\\ =0 & \text{if } \hat{x}_i^*\in(\xi_i^-,\xi_i^+),\\ \geq 0 & \text{if } \hat{x}_i^*=\xi_i^-,\end{cases} \quad i=1,2,\ldots,n,$$

which equals the following linear variational inequality [24, 28]: to find an $\hat{x}^*\in\Omega_1$ such that, for all $\hat{x}\in\Omega_1=\{\hat{x}|\xi^-\leq\hat{x}^*\leq\xi^+\}$,

$$\big(\hat{x}-\hat{x}^*\big)^{\mathrm{T}}\big(Q\hat{x}^*+\hat{q}-J^{\mathrm{T}}u^*+A^{\mathrm{T}}v^*\big)\geq 0. \tag{5.30}$$

Similarly, defining $\Omega_2 := \{v | v \geq 0\}$, we have the following LVI for (5.25) and (5.27): to find a $v^* \in \Omega_2$ such that

$$\left(v - v^*\right)^{\mathrm{T}}\left(-A\hat{x}^* + \hat{b}\right) \geq 0 \quad \forall v \in \Omega_2; \tag{5.31}$$

and the equality constraint (5.24) is equivalent to the following LVI: to find a $u^* \in \Omega_3 := \{u | u \in R^m\}$ such that

$$\left(u - u^*\right)^{\mathrm{T}}\left(J\hat{x}^* - \hat{d}\right) \geq 0 \quad \forall u \in \Omega_3. \tag{5.32}$$

Define $\Omega = \Omega_1 \times \Omega_2 \times \Omega_3 = \{\hat{y} := (\hat{x}^{\mathrm{T}}, u^{\mathrm{T}}, v^{\mathrm{T}})^{\mathrm{T}} \in R^{n+m+\dim(\hat{b})} | \xi^- \leq \hat{x}^* \leq \xi^+, u \text{ unrestricted}, v \geq 0\}$ [26, 28]. The linear variational inequalities (5.30)–(5.32) can be combined into one linear variational inequality problem formulation. That is, to find $\hat{y}^* := [\hat{x}^{*\mathrm{T}}, u^{*\mathrm{T}}, v^{*\mathrm{T}}]^{\mathrm{T}} \in \Omega$ such that, for all $\hat{y} = [\hat{x}^{\mathrm{T}}, u^{\mathrm{T}}, v^{\mathrm{T}}]^{\mathrm{T}} \in \Omega$, the following inequality always holds:

$$\left(\begin{bmatrix}\hat{x}\\ u\\ v\end{bmatrix} - \begin{bmatrix}\hat{x}^*\\ u^*\\ v^*\end{bmatrix}\right)^{\mathrm{T}}\left(\begin{bmatrix}Q & -J^{\mathrm{T}} & A^{\mathrm{T}}\\ J & 0 & 0\\ -A & 0 & 0\end{bmatrix}\begin{bmatrix}\hat{x}^*\\ u^*\\ v^*\end{bmatrix} + \begin{bmatrix}\hat{q}\\ -\hat{d}\\ \hat{b}\end{bmatrix}\right) \geq 0.$$

After defining $\varsigma^{\pm}$, H, and $\hat{p}$ respectively as in (5.12) and (5.13) for notation and implementation simplicity, the above LVI is exactly in the same compact matrix form as in (5.11), being the equivalence of QP (5.7)–(5.10). □

Theorem 5.3 *Assume the existence of optimal solution $\hat{x}^*$ to QP problem* (5.7)–(5.10). *LVI* (5.11) *is equivalent to the following piecewise linear equation* (*PLE*) (5.14).

Proof There are two steps to prove the equivalence of LVI (5.11) and PLE (5.14), including the proofs of sufficiency and necessity.

Step 1 (Sufficiency). Since $\hat{y}^* \in \Omega$ and $\hat{y} \in \Omega$ are guaranteed in LVI (5.11), there are three situations to be discussed as follows.

Situation 1. If $\hat{y}_i^* = \varsigma_i^+$, due to $\hat{y}_i - \hat{y}_i^* = \hat{y}_i - \varsigma_i^+ \leq 0$ and $(\hat{y}_i - \hat{y}_i^*)(H\hat{y}^* + \hat{p})_i \geq 0$, we have $(H\hat{y}^* + \hat{p})_i \leq 0$. Then $[\mathscr{P}_\Omega(\hat{y}^* - (H\hat{y}^* + \hat{p}))]_i = \varsigma_i^+ = \hat{y}_i^*$.

Situation 2. If $\varsigma_i^- < \hat{y}_i^* < \varsigma_i^+$, then $(\hat{y}_i - \hat{y}_i^*)$ is indefinite; while $(\hat{y}_i - \hat{y}_i^*)(H\hat{y}^* + \hat{p})_i \geq 0$ should always be true. So, $(H\hat{y}^* + \hat{p})_i = 0$, and we have $[\mathscr{P}_\Omega(\hat{y}^* - (H\hat{y}^* + \hat{p}))]_i = \hat{y}_i^*$.

Situation 3. If $\hat{y}_i^* = \varsigma_i^-$, due to $\hat{y}_i - \hat{y}_i^* = \hat{y}_i - \varsigma_i^- \geq 0$ and $(\hat{y}_i - \hat{y}_i^*)(H\hat{y}^* + \hat{p})_i \geq 0$, we have $(H\hat{y}^* + \hat{p})_i \geq 0$. Then $[\mathscr{P}_\Omega(\hat{y}^* - (H\hat{y}^* + \hat{p}))]_i = \varsigma_i^- = \hat{y}_i^*$.

In view of the above three situations, we always have

$$\left[\mathscr{P}_\Omega\left(\hat{y}_i^* - \left(H\hat{y}^* + \hat{p}\right)_i\right)\right]_i = \hat{y}_i^*$$

for any $i \in \{1, 2, \ldots, n + m + \dim(\hat{d})\}$. Therefore, if LVI (5.11) holds with $\hat{y}^*$, then $\hat{y}^*$ is also a solution of PLE (5.14).

Step 2 (Necessity). Consider that $\hat{y}^*$ is a solution of PLE (5.14). That is, $\hat{y}^* - \mathscr{P}_\Omega(\hat{y}^* - (H\hat{y}^* + \hat{p})) = 0$. As $\mathscr{P}_\Omega(\hat{y}^* - (H\hat{y}^* + \hat{p})) \in \Omega = \{\hat{y} \mid \varsigma^- \leq \hat{y} \leq \varsigma^+\}$, we have $\hat{y}^* \in \Omega$. There are thus three situations to be discussed as follows [with $i = 1, 2, \ldots, n + m + \dim(\hat{d})$].

Situation 1. If $\hat{y}_i^* = \varsigma_i^+$, due to $[\mathscr{P}_\Omega(\hat{y}^* - (H\hat{y}^* + \hat{p}))]_i - \hat{y}_i^* = 0$, i.e., $[\mathscr{P}_\Omega(\hat{y}^* - (H\hat{y}^* + \hat{p}))]_i = \varsigma_i^+$, we know $\varsigma_i^+ - (H\hat{y}^* + \hat{p})_i \geq \varsigma_i^+$. Thus, $(H\hat{y}^* + \hat{p})_i \leq 0$. For any $\hat{y} \in \Omega$ with $\varsigma_i^- \leq \hat{y}_i \leq \varsigma_i^+$, we have $\hat{y}_i - \varsigma_i^+ \leq 0$ and $(\hat{y}_i - \hat{y}_i^*)(H\hat{y}^* + \hat{p})_i = (\hat{y}_i - \varsigma_i^+)(H\hat{y}^* + \hat{p})_i \geq 0$.

Situation 2. If $\varsigma_i^- < \hat{y}_i^* < \varsigma_i^+$, due to $[\mathscr{P}_\Omega(\hat{y}^* - (H\hat{y}^* + \hat{p}))]_i - \hat{y}_i^* = 0$, we have $\hat{y}_i^* - (H\hat{y}^* + \hat{p})_i - \hat{y}_i^* = 0$, i.e., $(H\hat{y}^* + \hat{p})_i = 0$. So, for any $\hat{y} \in \Omega$ with $\varsigma_i^- \leq \hat{y}_i \leq \varsigma_i^+$, we have $(\hat{y}_i - \hat{y}_i^*)(H\hat{y}^* + \hat{p})_i = 0$, which makes the inequality $(\hat{y}_i - \hat{y}_i^*)(H\hat{y}^* + \hat{p})_i \geq 0$ hold true (by taking the equal sign).

Situation 3. If $\hat{y}_i^* = \varsigma_i^-$, due to $[\mathscr{P}_\Omega(\hat{y}^* - (H\hat{y}^* + \hat{p}))]_i - \hat{y}_i^* = 0$, i.e., $[\mathscr{P}_\Omega(\hat{y}^* - (H\hat{y}^* + \hat{p}))]_i = \varsigma_i^-$, we know that $\varsigma_i^- - (H\hat{y}^* + \hat{p})_i \leq \varsigma_i^-$. It follows that $(H\hat{y}^* + \hat{p})_i \geq 0$. For any $\hat{y} \in \Omega$ with $\varsigma_i^- \leq \hat{y}_i \leq \varsigma_i^+$, we have $\hat{y}_i - \varsigma_i^- \geq 0$ and $(\hat{y}_i - \hat{y}_i^*)(H\hat{y}^* + \hat{p})_i = (\hat{y}_i - \varsigma_i^-)(H\hat{y}^* + \hat{p})_i \geq 0$.

In view of the above three situations, we always have

$$(\hat{y}_i - \hat{y}_i^*)\big(H\hat{y}^* + \hat{p}\big)_i \geq 0$$

for any $i \in \{1, 2, \ldots, n + m + \dim(\hat{d})\}$ and $\hat{y}_i \in [\varsigma_i^-, \varsigma_i^+]$. Thus, $(\hat{y} - \hat{y}^*)^{\mathrm{T}}(H\hat{y}^* + \hat{p}) \geq 0$ holds; that is, the solution $\hat{y}^*$ of PLE (5.14) also satisfies LVI (5.11).

In summary, Steps 1 and 2 collectively prove the equivalence of LVI (5.11) and PLE (5.14); i.e., the sufficiency and necessity of the second conversion. □

5.2.4 *Numerical Studies*

Theoretical results about LVI-PDNN (5.15) depicted in previous subsections are substantiated and supplemented by the following numerical observations. The first one is about the global convergence and exponential convergence. The second one is about the solution behavior starting from initial states within feasible region. Finally, the third one is about the convergence of the LVI-PDNN under the circumstance of no $\hat{x}^*$ solution.

5.2.4.1 Global Convergence

The PDNN model (5.15) for solving QP/LP (5.7)–(5.10) is simulated with randomly generated coefficient matrices and vectors, randomly generated initial states $\hat{y}(0)$, and without loss of generality, $n = 7$, $m = 3$, $\dim(\hat{b}) = 1$, $\bar{\gamma} = 10^4$. Assuming the existence of $\hat{x}^*$, the global convergence of LVI-PDNN (5.15) is illustrated in Fig. 5.10 wherein the usual convergence time is less than 4×10^{-3}

second. As shown in Fig. 5.11, the exponential-convergence factor $\tilde{\varrho}$ is estimated online as $\|\hat{y} - \mathscr{P}_\Omega(\hat{y} - (H\hat{y} + \hat{p}))\|_2^2/\|\hat{y} - \hat{y}^*\|_2^2$ for all $\hat{y} \neq \hat{y}^*$, and otherwise, $\lim_{\hat{y}\to\hat{y}^*} \|\hat{y} - \mathscr{P}_\Omega(\hat{y} - (H\hat{y} + \hat{p}))\|_2^2/\|\hat{y} - \hat{y}^*\|_2^2$. The value of ϱ in the proof of Theorem 5.1 can thus be chosen as $\min(\tilde{\varrho}(t))$. This typical simulation result justifies the exponential convergence property of the LVI-based PDNN model. Furthermore, the $\hat{x}^*$ values are also compared with those by MATLAB QP routines: the difference is less than 10^{-7}.

5.2.4.2 Feasible-Region Solutions

As in the robotic applications, the initial state $\hat{x}(0)$ (being the initial joint velocity, acceleration, or torque) is usually within the feasible region constituted by constraints (5.7)–(5.10). Thus, we may be interested in the question whether the network output $\hat{x}(t)$ starting from such an $\hat{x}(0)$ will always be within the same feasible region. If not, what measurements could we adopt? Figure 5.12 shows the typical situation wherein only $\varphi = 0$ corresponds to $\hat{x}(t)$ being within the feasible region. From the figure we see that $\hat{x}(t)$ sometimes does go out of the region. In light of the exponential convergence, this weakness can be remedied easily by increasing the design parameter $\bar{\gamma}$ as large as the hardware would permit (e.g., $\bar{\gamma} = 10^7$, so that the shifting period is only of level 10^{-7}second) or using a limiter to force the network output $\hat{x}(t)$ between $[\xi^-, \xi^+]$. This numerical result also applies to other neural networks design for robot manipulators and might be the source of the small end-effector position error of level 10^{-4}–10^{-7} m in [3, 15–20, 23, 25, 29].

5.2.4.3 No-Solution Case

Another interesting topic of using the LVI-PDNN (5.15) is about the case of no optimal/theoretical solutions $\hat{x}^*$ to the original QP/LP problem (5.7)–(5.10). If so, what convergence behavior will the network illustrate? The nonexistence of $\hat{x}^*$ actually means the feasible region being empty. In robotics, this is an extreme case but does happen, like to command the robot arm to do an impossible task of positioning or lifting. Though the theoretical result does not cover this case, a large number of random numerical tests show that the network output $\hat{x}(t)$ is always convergent (while the dual variables u and v are not). Due to space limitation, there are no figures to show. The typical $\hat{x}(t)$ convergence is similar to Fig. 5.10. This numerical result is also applicable to other neural networks design for robot manipulators and may explain why no divergence of joint variables was observed even in the extreme case (instead, a jump or loss of integration precision may occur) [3, 15–20, 23, 25, 29].

5.2.5 LVI-PDNN for Solving RMP Scheme

Considering the applications on repetitive motion planning and control of redundant robot manipulators, we develop the specific LVI-based primal–dual neural network

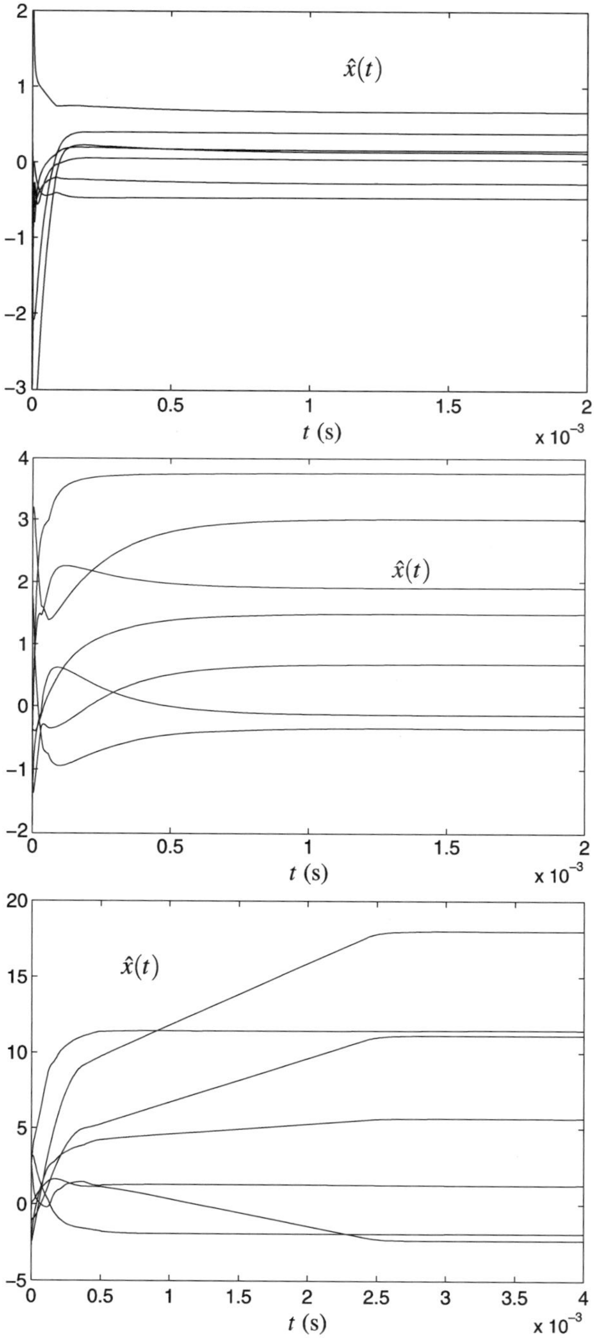

Fig. 5.10 Global and exponential convergence of LVI-PDNN (5.15) starting from randomly generated initial states to $\hat{x}^*$ of QP (positive definite), QP (positive semi-definite), and LP, corresponding to the *upper*, *middle*, and *lower graphs*, respectively

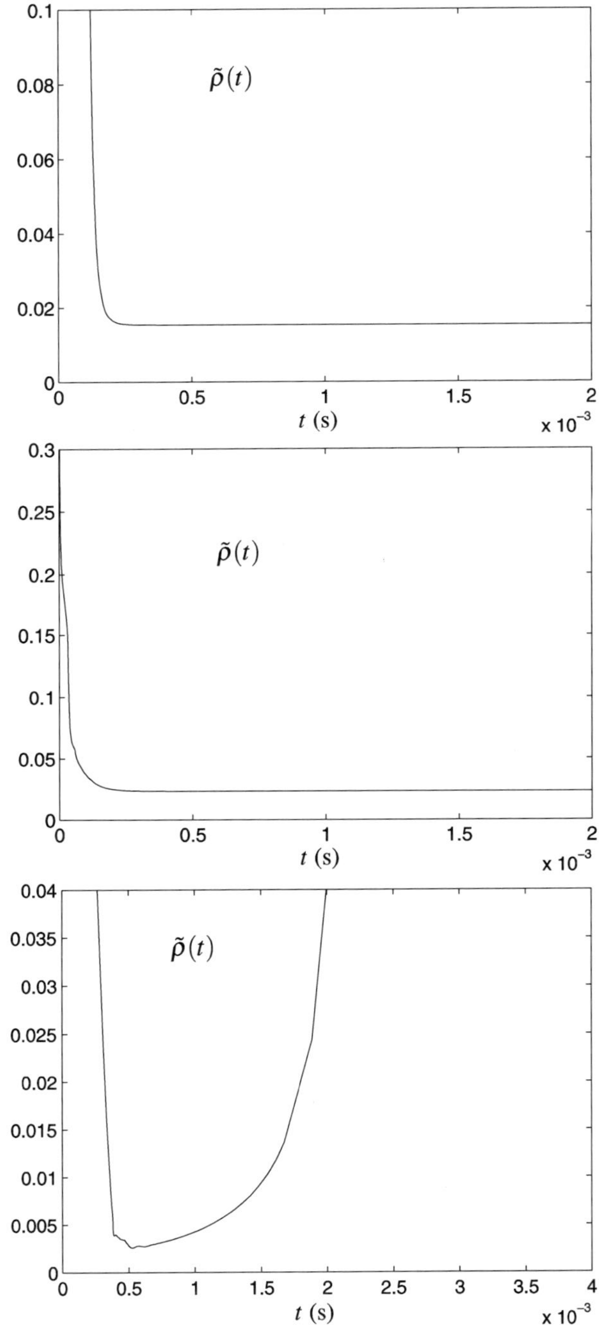

Fig. 5.11 Exponential-convergence factor $\tilde{\varrho}(t)$ of LVI-PDNN (5.15) starting from randomly generated initial states for QP (positive definite), QP (positive semi-definite), and LP, corresponding to the *upper*, *middle*, and *lower graphs*, respectively

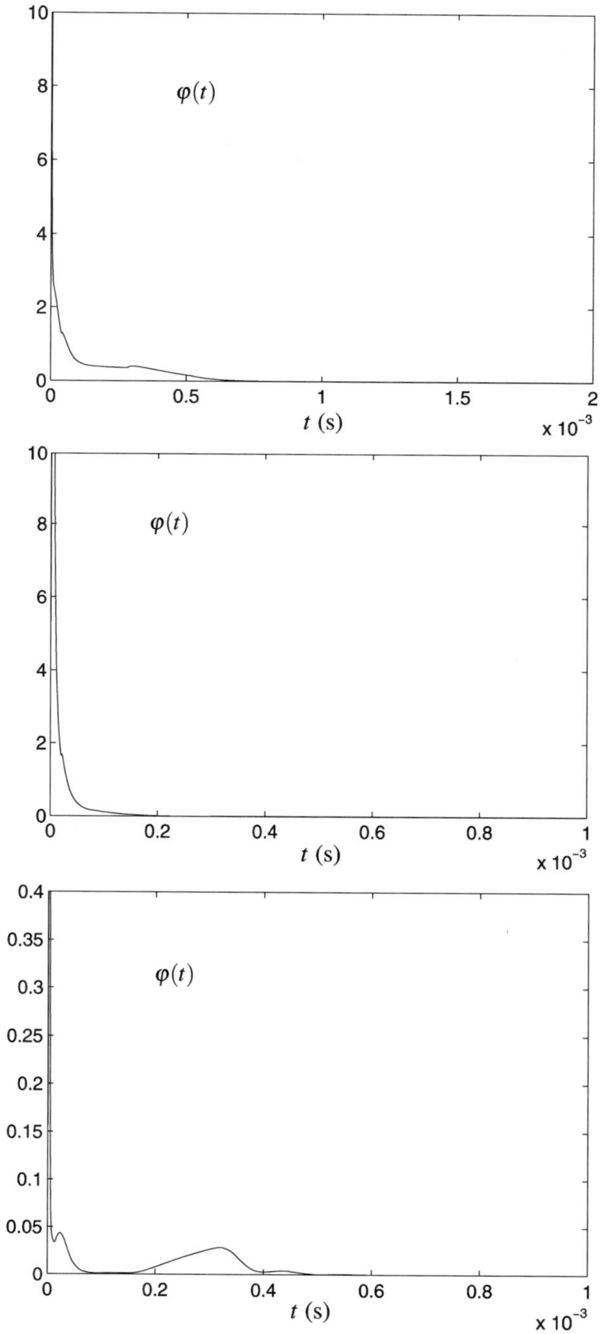

Fig. 5.12 The solution processes of QP (positive definite), QP (positive semi-definite), and LP may shift out of the feasible region, though started from it, which correspond to the *upper*, *middle*, and *lower graphs*, respectively

in this subsection for solving the QP problem (2.22)–(2.24), which is only subject to equality and bound constraints. The specific LVI-PDNN design procedure is the same as before but repeated as follows.

First, we convert the QP problem (2.22)–(2.24) to a linear variational inequality. That is, to find a primal–dual equilibrium vector $\hat{y}^* \in \Omega := \{\hat{y} \mid \varsigma^- \leq \hat{y} \leq \varsigma^+\} \subset R^{n+m}$ such that

$$(\hat{y} - \hat{y}^*)^{\mathrm{T}}(H\hat{y}^* + \hat{p}) \geq 0 \quad \forall \hat{y} \in \Omega, \tag{5.33}$$

where the primal–dual decision variable vector $\hat{y}$ and its upper/lower bounds $\varsigma^{\pm}$ (in this LVI-based primal–dual neural-network design) are defined respectively as (with $\varpi \gg 0 \in R^m$)

$$\hat{y} = \begin{bmatrix} \hat{x} \\ u \end{bmatrix}, \qquad \varsigma^+ = \begin{bmatrix} \xi^+ \\ \varpi \end{bmatrix}, \qquad \varsigma^- = \begin{bmatrix} \xi^- \\ -\varpi \end{bmatrix} \in R^{m+n}.$$

In (5.33), the coefficient matrix H and vector $\hat{p}$ are defined as follows:

$$H = \begin{bmatrix} Q & -J^{\mathrm{T}} \\ J & 0 \end{bmatrix}, \qquad \hat{p} = \begin{bmatrix} \hat{q} \\ -\hat{d} \end{bmatrix} \in R^{n+m}.$$

Theoretically, we have the following statement of this design step.

Theorem 5.4 *By assuming the existence of optimal solution $\hat{x}^*$ [in view of $Q = I$ in the most preferred RMP scheme* (2.2)–(2.5)], *the QP problem* (2.22)–(2.24) *can be converted to the LVI problem* (5.33).

It is known that the LVI problem (5.33) is equivalent to the piecewise-linear equation

$$\mathscr{P}_\Omega\big(\hat{y} - (H\hat{y} + \hat{p})\big) - \hat{y} = 0, \tag{5.34}$$

where $\mathscr{P}_\Omega(\cdot)$ is the projection operator from space R^{n+m} onto the set Ω.

To solve LVI (5.34) and QP (2.22)–(2.24), guided by dynamic-system-solver design experience [15], we could adopt the following dynamic equation of the so-called LVI-based primal–dual neural network:

$$\dot{\hat{y}} = \bar{\gamma}(I + H^{\mathrm{T}})\big(\mathscr{P}_\Omega\big(\hat{y} - (H\hat{y} + \hat{p})\big) - \hat{y}\big), \tag{5.35}$$

where the design parameter $\bar{\gamma} > 0$ is defined to be the same as that for dual neural network in Sect. 4.5. Furthermore, we have the following important results [15].

Theorem 5.5 *Assume the existence of optimal solution $\hat{x}^*$ to QP* (2.22)–(2.24) (*in view of $Q = I$). Starting from any initial state $\hat{y}(0)$, the LVI-based primal–dual neural network* (5.35) *converges to an equilibrium point $\hat{y}^*$, of which the first n elements constitute the optimal solution $\hat{x}^*$ to the inverse-kinematics QP problem* (2.22)–(2.24). *Moreover, if there exists a constant $\rho > 0$ such that $\|\hat{y} - \mathscr{P}_\Omega(\hat{y} - (H\hat{y} + \hat{p}))\|_2^2 \geq \rho\|\hat{y} - \hat{y}^*\|_2^2$, then the exponential convergence can be achieved for such an LVI-based primal–dual neural network. The exponential convergence rate is proportional to $\bar{\gamma}\rho$.*

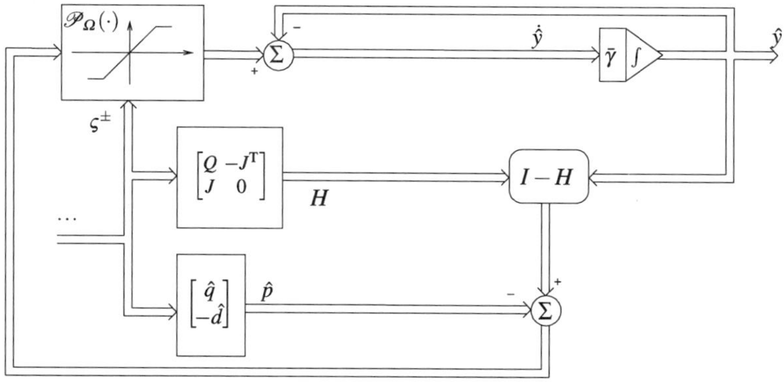

Fig. 5.13 Block diagram of simplified LVI-based primal–dual neural network (5.36)

5.3 Simplified LVI-Based Primal–Dual Neural Network

By removing the scaling term $(I + M^{\mathrm{T}})$ of LVI-PDNN (5.35), we get the following simplified LVI-based primal–dual neural network:

$$\dot{\hat{y}} = \bar{\gamma}\big(\mathscr{P}_\Omega\big(\hat{y} - (H\hat{y} + \hat{p})\big) - \hat{y}\big), \tag{5.36}$$

where $\bar{\gamma} \in R$ is still a positive design parameter used to adjust the convergence rate of the neural networks. The block diagram realizing (5.36) is presented in Fig. 5.13. Note that, according to the definition of $\hat{y}$, being the first n elements of $\dot{\hat{y}}$ in (5.36), the joint acceleration $\ddot{\theta}$ can also be explicitly generated for joint torque control [15] if needed. Furthermore, we have the following theoretical results about the simplified LVI-PDNN (5.36) [15, 30].

Theorem 5.6 *Assume the existence of optimal solution $\hat{x}^*$ to the strictly convex QP problem depicted in* (2.22)–(2.24). *The output $\hat{x}(t)$ of the simplified LVI-based primal-dual neural network* (5.36) *globally exponentially converges to the optimal solution $\hat{x}^*$. In addition, the exponential-convergence rate is proportional to the product of $\bar{\gamma}$ and the minimum eigenvalue of Q (being* 1 *in our RMP case).*

Corollary 5.1 (Existence of $\hat{y}^*$) *The existence of an equilibrium point $\hat{y}^*$ of the simplified LVI-PDNN model* (5.36) *is equivalent to the nonemptiness of the feasible region made by constraints* (2.23) *and* (2.24), *denoted by $\Phi(\hat{x})$.*

Corollary 5.2 (Nonexistence of $\hat{y}^*$) *If the feasible region made by constraints* (2.23) *and* (2.24) *is empty* [*i.e., $\Phi(\hat{x}) = \phi$*], *then the state $\hat{y}(t)$ of the simplified LVI-PDNN model* (5.36) *is generally divergent.*

Corollary 5.3 (General QP Solution) *Assume the existence of $\hat{y}^*$ as in Corollary* 5.1 [*i.e., $\Phi(\hat{x}) \neq \phi$*]. *For solving general QP problems of Q being positive semidefinite or even zero (i.e., the LP situation), the output $\hat{x}(t)$ of the simplified LVI-based primal–dual neural network* (5.36) *may be divergent or oscillatory.*

The nonexistence of $\hat{y}^*$ (or $\hat{x}^*$) means that the feasible region $\Phi(\hat{x})$ is empty. Corollaries 5.1 and 5.2 give the general divergence result for this no-solution case. However, a large number of randomly generated numerical tests show that, although the state $\hat{y}$ is usually divergent, the network output $\hat{x}(t)$ of simplified LVI-PDNN (5.36) is always convergent [30]. This property could be very useful in the robotic applications.

5.4 Chapter Conclusions

In this chapter, the rules for deriving the dual problems of linear programs have been first investigated and applied to the development of a traditional primal–dual neural network. The neural network could solve an online linear program and its dual problem simultaneously in the context of multiple solutions, unbounded solutions, and/or unique solution. In addition, the LVI-based primal–dual neural network has been investigated thoroughly for solving the general QP/LP problems. Furthermore, a simplified LVI-based primal–dual neural network has been derived by removing the scaling matrix. Theoretical analysis and numerical results have substantiated the efficacy of such three neural networks.

References

1. Zhang Y, Yi C, Ma W (2010) On a primal–dual neural network for online solution of linear programming. OR Trans 14(3):1–10
2. Xia Y (1996) A new neural network for solving linear programming problems and its applications. IEEE Trans Neural Netw 7(2):525–529
3. Tang WS, Wang J (2001) A recurrent neural network for minimum infinity-norm kinematic control of redundant manipulators with an improved problem formulation and reduced architecture complexity. IEEE Trans Syst Man Cybern, Part B, Cybern 31(1):98–105
4. Wang J, Zhang Y (2004) Recurrent neural networks for real-time computation of inverse kinematics of redundant manipulators. In: Sincak P, Vascak J, Hirota K (eds) Machine intelligence: quo vadis? World Scientific, Singapore, pp 299–319
5. Zhang Y (2006) Towards piecewise-linear primal neural networks for optimization and redundant robotics. In: Proceedings of IEEE international conference on networking, sensing and control, pp 374–379
6. Ham FM, Kostanic I (2001) Principles of neurocomputing for science and engineering. McGraw-Hill, New York
7. Zhang Y, Leithead WE, Leith DJ (2005) Time-series Gaussian process regression based on Toeplitz computation of $O(N^2)$ operations and $O(N)$-level storage. In: Proceedings of the 44th IEEE conference on decision and control, pp 3711–3716
8. Pyne IB (1956) Linear programming on an electronic analogue computer. Trans Am Inst Electr Eng 75:139–143
9. Mead C (1989) Analog VLSI and neural systems. Addison-Wesley, Reading
10. Zhang Y (2007) Dual neural networks: design, analysis, and application to redundant robotics. In: Kang GB (ed) Progress in neurocomputing research. Nova Science, New York, pp 41–81
11. Bazaraa MS, Sherali HD, Shetty CM (1993) Nonlinear programming—theory and algorithms. Wiley, New York

12. The MathWorks, Inc (2007) Using Simulink. Natick, MA
13. Zhang Y (2005) On the LVI-based primal–dual neural network for solving online linear and quadratic programming problems. In: Proceedings of American control conference, pp 1351–1356
14. Zhang Y, Wang J (2002) A dual neural network for convex quadratic programming subject to linear equality and inequality constraints. Phys Lett A 298(4):271–278
15. Zhang Y, Ge SS, Lee TH (2004) A unified quadratic programming based dynamical system approach to joint torque optimization of physically constrained redundant manipulators. IEEE Trans Syst Man Cybern, Part B, Cybern 34(5):2126–2132
16. Zhang Y, Wang J (2002) A dual neural network for constrained torque optimization of kinematically redundant manipulators. IEEE Trans Syst Man Cybern, Part B, Cybern 32(5):654–662
17. Zhang Y, Wang J, Xu Y (2002) A dual neural network for bi-criteria kinematic control of redundant manipulators. IEEE Trans Robot Autom 18(6):923–931
18. Zhang Y, Wang J, Xia Y (2003) A dual neural network for redundancy resolution of kinematically redundant manipulators subject to joint limits and joint velocity limits. IEEE Trans Neural Netw 14(3):658–667
19. Zhang Y, Wang J (2004) Obstacle avoidance of kinematically redundant manipulators using a dual neural network. IEEE Trans Syst Man Cybern, Part B, Cybern 34(1):752–759
20. Ge SS, Zhang Y, Lee TH (2004) An acceleration-based weighting scheme for minimum-effort inverse kinematics of redundant manipulators. In: Proceedings of IEEE international symposium on intelligent control, pp 275–280
21. Kennedy MP, Chua LO (1988) Neural networks for nonlinear programming. IEEE Trans Circuits Syst 35(5):554–562
22. Zhang S, Constantinides AG (1992) Lagrange programming neural networks. IEEE Trans Circuits Syst 39(7):441–452
23. Zhang Y, Wang J (2003) Obstacle avoidance of redundant manipulators using a dual neural network. In: Proceedings of IEEE international conference on robotics and automation, pp 2747–2752
24. Xia Y, Wang J (2000) A recurrent neural network for solving linear projection equations. Neural Netw 13(3):337–350
25. Zhang Y, Ge SS (2003) A general recurrent neural network model for time-varying matrix inversion. In: Proceedings of the 42nd IEEE conference on decision and control, pp 6169–6174
26. He B (1992) A projection and contraction method for a class of linear complementarity problems and its application in convex quadratic programming. Appl Math Optim 25(3):247–262
27. Zhang Y, Guo D, Li K, Li J (2012) Manipulability-maximizing self-motion planning and control of redundant manipulators with experimental validation. In: Proceedings of IEEE international conference on mechatronics and automation, pp 1829–1834
28. Kinderlehrer D, Stampcchia G (1980) An introduction to variational inequalities and their applications. Academic Press, New York
29. Cheng FT, Chen TH, Sun YY (1994) Resolving manipulator redundancy under inequality constraints. IEEE J Robot Autom 10(1):65–71
30. Zhang Y, Li Z, Tan H, Fan Z (2007) On the simplified LVI-based primal–dual neural network for solving LP and QP problems. In: Proceedings of the 6th IEEE international conference on control and automation, pp 3129–3134

Chapter 6
Numerical Algorithm 94LVI

Abstract This chapter presents and investigates a numerical algorithm (termed 94LVI algorithm) for solving the QP problem subject to linear equality and bound constraints. To do this, as the previous chapter shows, the constrained QP problem is first converted into the LVI, which is then converted into an equivalent piecewise-linear equation (PLE). After that, the resultant PLE is solved by the presented 94LVI algorithm. The optimal numerical solution to the QP problem is thus obtained. Furthermore, the theoretical proof of the global convergence of the 94LVI algorithm is presented. Moreover, the numerical comparison results between the 94LVI algorithm and the active-set algorithm are provided, which further demonstrate the efficacy and superiority of the presented 94LVI algorithm for solving the QP problem.

6.1 Introduction

The online solution of the QP problem is widely encountered in various areas, e.g., optimal controller design [1], power-scheduling [2], robot-arm motion planning [3], and digital signal processing [4]. Motivated by engineering applications of QP in robotics [5–7], especially the RMP application, the following problem formulation is considered in this chapter and mostly in this book:

$$\begin{aligned} &\text{minimize } \hat{x}^{\mathrm{T}} Q\hat{x}/2 + \hat{q}^{\mathrm{T}}\hat{x},\\ &\text{subject to } J\hat{x} = \hat{d},\\ &\qquad \xi^- \le \hat{x} \le \xi^+, \end{aligned} \tag{6.1}$$

where $\hat{x} \in R^n$ is the decision vector to be obtained, and $Q \in R^{n\times n}$ is a positive definite symmetric matrix. The other coefficient matrices and vectors are defined respectively as $\hat{q} \in R^n$, $J \in R^{m\times n}$, and $\hat{d} \in R^m$. The n-dimensional vectors ξ^- and ξ^+ denote respectively the lower and upper bounds of $\hat{x}$.

In the past decades, a large number of numerical algorithms have been proposed and developed for solving constrained optimization problems in the form of QP (6.1), such as the Lagrangian multiplier method [8, 9], active-set method [10],

Y. Zhang, Z. Zhang, *Repetitive Motion Planning and Control of Redundant Robot Manipulators*, DOI 10.1007/978-3-642-37518-7_6,

weighted-path-following interior-point algorithm [11], and projection and contraction algorithm [12]. Note that most of the numerical algorithms are based on the active-set strategy, and MATLAB function "quadprog" integrates such an active-set method. In this chapter [13], as the previous chapter shows, such a QP problem is first converted to the linear variational inequality (LVI), which is equivalent to a piecewise-linear equation (PLE). Second, a numerical algorithm (termed 94LVI algorithm) is employed for solving the resultant PLE, and the optimal numerical solution to the QP problem subject to linear equality and bound constraints is thus obtained. Note that this work extends the paper [14], which was published in 1994 to solve the LVI problem, and, for simplicity, the presented numerical algorithm is named the 94LVI algorithm. Numerical-experiment results substantiate the efficacy and accuracy of the 94LVI algorithm. Compared with the active-set method, the 94LVI algorithm shortens the running time of the algorithm program for solving the QP problem observably.

The remainder of this chapter is organized into the following sections. The 94LVI algorithm for solving the QP problem subject to linear equality and bound constraints is presented and discussed in Sect. 6.2. The global convergence of the 94LVI algorithm is proved in Sect. 6.3. The efficacy and accuracy of the 94LVI algorithm are further investigated via numerical experiments in Sect. 6.4. Finally, Sect. 6.5 concludes this chapter with final remarks.

6.2 The 94LVI Algorithm for QP Solving

In the section, following the authors' previous design methods [15, 16], an LVI-based numerical algorithm for solving QP (6.1) is developed.

First, we convert QP (6.1) into the LVI problem, i.e., to find a vector $\hat{y}^* \in \Omega := \{\hat{y} \mid \varsigma^- \le \hat{y} \le \varsigma^+\} \subset R^{n+m}$ such that

$$(\hat{y} - \hat{y}^*)^{\mathrm{T}}(H\hat{y}^* + \hat{p}) \ge 0 \quad \forall \hat{y} \in \Omega, \tag{6.2}$$

where the primal–dual decision vector $\hat{y}$ and its upper and lower bounds $\varsigma^{\pm}$ are defined respectively as

$$\hat{y} = \begin{bmatrix} \hat{x} \\ u \end{bmatrix}, \quad \varsigma^+ = \begin{bmatrix} \xi^+ \\ \varpi \end{bmatrix}, \quad \varsigma^- = \begin{bmatrix} \xi^- \\ -\varpi \end{bmatrix}, \tag{6.3}$$

with $\varpi \gg 0 \in R^m$ defined again to replace the m-dimensional $+\infty$ numerically. In addition, the vector $\hat{x}$ is the primal/original decision vector of QP (6.1), and the vector u is the dual decision vector defined corresponding to the equality constraint in QP (6.1). The coefficients in LVI (6.2) are defined as

$$H = \begin{bmatrix} Q & -J^{\mathrm{T}} \\ J & 0 \end{bmatrix}, \quad \hat{p} = \begin{bmatrix} \hat{q} \\ -\hat{d} \end{bmatrix}.$$

According to [14, 15], LVI (6.2) is equivalent to the following piecewise-linear equation (PLE):

$$\mathscr{P}_\Omega\big(\hat{y} - (H\hat{y} + \hat{p})\big) - \hat{y} = 0, \tag{6.4}$$

where, as also defined and used in Chaps. 4 and 5, $\mathscr{P}_\Omega(\cdot) : R^{n+m} \to \Omega$ is a piecewise-linear projection operator with the ith element of $\mathscr{P}_\Omega(\hat{y})$ being

$$\big[\mathscr{P}_\Omega(\hat{y})\big]_i = \begin{cases} \varsigma_i^- & \text{if } \hat{y}_i < \varsigma_i^-, \\ \hat{y}_i & \text{if } \varsigma_i^- \le \hat{y}_i \le \varsigma_i^+, \\ \varsigma_i^+ & \text{if } \hat{y}_i > \varsigma_i^+, \end{cases} \quad i = 1, 2, \ldots, n+m.$$

Second, we can define a set Ω^* to denote the solution set of LVI (6.2) and PLE (6.4). To solve the problem, we can define the following vector-valued error function:

$$e(\hat{y}) = \hat{y} - \mathscr{P}_\Omega\big(\hat{y} - (H\hat{y} + \hat{p})\big), \tag{6.5}$$

so that solving PLE (6.4) is equivalent to finding a zero point of (6.5). As proposed in [14], let us define the state vector in the κth iteration as $\hat{y}^\kappa = [(\hat{x}^\kappa)^{\mathrm{T}}, (u^\kappa)^{\mathrm{T}}]^{\mathrm{T}}$. Given the initial state $\hat{y}^0 \in R^{n+m}$, for iteration index $\kappa = 0, 1, 2, \ldots$ (used as the superscript, e.g., superscript ^{0} in $\hat{y}^0$), if $\hat{y}^\kappa \notin \Omega^*$, then we can employ the following 94LVI iteration formula for solving (6.4):

$$\hat{y}^{\kappa+1} = \hat{y}^\kappa - \hat{\rho}(\hat{y}^\kappa)\sigma(\hat{y}^\kappa), \tag{6.6}$$

where $\sigma(\hat{y}^\kappa) := (H^{\mathrm{T}} + I)e(\hat{y}^\kappa)$ and $\hat{\rho}(\hat{y}^\kappa) := \|e(\hat{y}^\kappa)\|_2^2 / \|\sigma(\hat{y}^\kappa)\|_2^2$.

Third, for algorithm implementation purposes, we have the two-norm-based scalar-valued function

$$\big\|e(\hat{y}^\kappa)\big\|_2 = \big\|\hat{y}^\kappa - \mathscr{P}_\Omega\big(\hat{y}^\kappa - (Hy^\kappa + \hat{p})\big)\big\|_2, \tag{6.7}$$

and the control condition of the 94LVI algorithm implementation, which decides the accuracy of the solution, can be described as "in every iteration step of 94LVI algorithm, if the value of $\|e(\hat{y}^\kappa)\|_2$ is less than the preset accuracy value (PAV), the iteration stops. The value of $\hat{y}^\kappa$ that satisfies the PAV in the (current) κth iteration step is regarded as a solution $\hat{y}^*$ of PLE (6.4)." Note that the first n elements of $\hat{y}^*$ constitute the solution $\hat{x}^*$ to the original QP (6.1). It is worth mentioning that we also record the running time of the program and solution-errors in the program of the 94LVI algorithm, which are favorable for further demonstrating the efficacy and accuracy of the 94LVI algorithm.

6.3 Global Convergence

Following [14, 17, 18], the important lemmas about the convergence of the 94LVI algorithm are cited and presented below (with themes similar to those in [14] but with proof details given according to the authors' engineering-type understanding, in addition to the consideration on completeness of results and readers' convenience).

Lemma 6.1 *There exists* $\hat{y}^* \in \Omega^*$ *such that the sequence* $\{\hat{y}^\kappa\}$ *(with iteration index* $\kappa = 0, 1, 2, \ldots$*) generated by the* 94*LVI algorithm satisfies* $\|\hat{y}^{\kappa+1} - \hat{y}^*\|_2^2 \le \|\hat{y}^\kappa - \hat{y}^*\|_2^2 - \hat{\rho}(\hat{y}^\kappa)\|e(\hat{y}^\kappa)\|_2^2$.

Proof Referring to [14], using (6.6), we obtain

$$\begin{aligned}
\|\hat{y}^{\kappa+1} - \hat{y}^*\|_2^2 &= \|\hat{y}^\kappa - \hat{y}^* - \hat{\rho}(\hat{y}^\kappa)\sigma(\hat{y}^\kappa)\|_2^2 \\
&= \big[(\hat{y}^\kappa - \hat{y}^*) - \hat{\rho}(\hat{y}^\kappa)\sigma(\hat{y}^\kappa)\big]^{\mathrm{T}}\big[(\hat{y}^\kappa - \hat{y}^*) - \hat{\rho}(\hat{y}^\kappa)\sigma(\hat{y}^\kappa)\big] \\
&= \|\hat{y}^\kappa - \hat{y}^*\|_2^2 - (\hat{y}^\kappa - \hat{y}^*)^{\mathrm{T}}\hat{\rho}(\hat{y}^\kappa)\sigma(\hat{y}^\kappa) - \sigma^{\mathrm{T}}(\hat{y}^\kappa)\hat{\rho}(\hat{y}^\kappa)(\hat{y}^\kappa - \hat{y}^*) \\
&\quad + \sigma^{\mathrm{T}}(\hat{y}^\kappa)\hat{\rho}(\hat{y}^\kappa)\hat{\rho}(\hat{y}^\kappa)\sigma(\hat{y}^\kappa) \\
&= \|\hat{y}^\kappa - \hat{y}^*\|_2^2 - 2\hat{\rho}(\hat{y}^\kappa)(\hat{y}^\kappa - \hat{y}^*)^{\mathrm{T}}\sigma(\hat{y}^\kappa) + \hat{\rho}^2(\hat{y}^k)\|\sigma(\hat{y}^\kappa)\|_2^2 \\
&= \|\hat{y}^\kappa - \hat{y}^*\|_2^2 - 2\hat{\rho}(\hat{y}^\kappa)(\hat{y}^\kappa - \hat{y}^*)^{\mathrm{T}}(H^{\mathrm{T}} + I)e(\hat{y}^\kappa) \\
&\quad + \hat{\rho}(\hat{y}^\kappa)\|e(\hat{y}^\kappa)\|_2^2.
\end{aligned}$$

By Theorem 1 of [14], $(\hat{y}^\kappa - \hat{y}^*)^{\mathrm{T}}(H^{\mathrm{T}} + I)e(\hat{y}^\kappa) \ge \|e(\hat{y}^\kappa)\|^2 \ge 0$, whence we obtain

$$\begin{aligned}
\|\hat{y}^{\kappa+1} - \hat{y}^*\|_2^2 &\le \|\hat{y}^\kappa - \hat{y}^*\|_2^2 - 2\hat{\rho}(\hat{y}^\kappa)\|e(\hat{y}^\kappa)\|_2^2 + \hat{\rho}(\hat{y}^\kappa)\|e(\hat{y}^\kappa)\|_2^2 \\
&= \|\hat{y}^\kappa - \hat{y}^*\|_2^2 - \hat{\rho}(\hat{y}^\kappa)\|e(\hat{y}^\kappa)\|_2^2.
\end{aligned} \tag{6.8}$$

The proof is thus completed. □

Lemma 6.2 *The sequence* $\{\hat{y}^\kappa\}$ *(with iteration index* $\kappa = 0, 1, 2, \ldots$*) generated by the* 94*LVI algorithm converges to a solution* $\hat{y}^*$*. In addition, the first n elements of* $\hat{y}^*$ *constitute the optimal solution* $\hat{x}^*$ *to QP* (6.1).

Proof From (6.6) we get

$$\begin{aligned}
\hat{\rho}(\hat{y}^\kappa) &= \|e(\hat{y}^\kappa)\|_2^2 / \|\sigma(\hat{y}^\kappa)\|_2^2 \\
&= \|e(\hat{y}^\kappa)\|_2^2 / \|(H^{\mathrm{T}} + I)e(\hat{y}^\kappa)\|_2^2 \\
&\ge \|e(\hat{y}^\kappa)\|_2^2 / \big(\|H^{\mathrm{T}} + I\|_{\mathrm{F}}^2\|e(\hat{y}^\kappa)\|_2^2\big) \\
&= 1/\|H^{\mathrm{T}} + I\|_{\mathrm{F}}^2 > 0,
\end{aligned} \tag{6.9}$$

where the symbol $\|\cdot\|_{\mathrm{F}}$ denotes the Frobenius norm of a matrix. Combining inequalities (6.8) and (6.9), we obtain

$$\|\hat{y}^{\kappa+1}-\hat{y}^*\|_2^2 \le \|\hat{y}^{\kappa}-\hat{y}^*\|_2^2 - \|e(\hat{y}^{\kappa})\|_2^2/\|H^{\mathrm{T}}+I\|_{\mathrm{F}}^2 \le \|\hat{y}^{\kappa}-\hat{y}^*\|_2^2, \quad (6.10)$$

from which we get the following inequality:

$$\|e(\hat{y}^{\kappa})\|_2^2/\|H^{\mathrm{T}}+I\|_{\mathrm{F}}^2 \le \|\hat{y}^{\kappa}-\hat{y}^*\|_2^2 - \|\hat{y}^{\kappa+1}-\hat{y}^*\|_2^2.$$

By defining $1/\|H^{\mathrm{T}}+I\|_{\mathrm{F}}^2$ as δ, for iteration indices $\kappa = 0, 1, 2, \ldots$, we further obtain

$$\delta\sum_{\kappa=0}^{+\infty}\|e(\hat{y}^{\kappa})\|_2^2 \le \|\hat{y}^0-\hat{y}^*\|_2^2 - \lim_{\kappa\to+\infty}\|\hat{y}^{\kappa+1}-\hat{y}^*\|_2^2 \le \|\hat{y}^0-\hat{y}^*\|_2^2, \quad (6.11)$$

and, mathematically, there must exist a positive number h satisfying $0 \le h \le \|\hat{y}^0-\hat{y}^*\|_2^2$ and

$$\lim_{j\to\infty}\delta\sum_{\kappa=0}^{j}\|e(\hat{y}^{\kappa})\|_2^2 = h.$$

Then, according to the necessary condition of series convergence (i.e., a corollary from the Cauchy criterion of series convergence), from (6.11) we obtain

$$\lim_{\kappa\to\infty}\|e(\hat{y}^{\kappa})\|_2^2 = 0.$$

By defining $e(\hat{y}^{\kappa}) := [e_1(\hat{y}^{\kappa}), e_2(\hat{y}^{\kappa}), \ldots, e_{m+n}(\hat{y}^{\kappa})]^{\mathrm{T}}$, we then have

$$\lim_{\kappa\to\infty}\|e(\hat{y}^{\kappa})\|_2^2 = \lim_{\kappa\to\infty}\left[\left(e_1(\hat{y}^{\kappa})\right)^2 + \left(e_2(\hat{y}^{\kappa})\right)^2 + \cdots + \left(e_{m+n}(\hat{y}^{\kappa})\right)^2\right] = 0,$$

from which it follows that

$$\lim_{\kappa\to\infty} e_1(\hat{y}^{\kappa}) = 0,\ \lim_{\kappa\to\infty} e_2(\hat{y}^{\kappa}) = 0, \ldots, \lim_{\kappa\to\infty} e_{m+n}(\hat{y}^{\kappa}) = 0. \quad (6.12)$$

From (6.12) we further obtain

$$\lim_{\kappa\to\infty} e(\hat{y}^{\kappa}) = 0.$$

Let $\hat{y}^*$ be a solution of PLE (6.4). Then the sequence $\{\hat{y}^{\kappa}\}$ has exactly a cluster point, and

$$\lim_{\kappa\to\infty}\hat{y}^{\kappa} = \hat{y}^* \quad \text{with } e(\hat{y}^*) = 0.$$

In addition, the first n elements of $\hat{y}^*$ (i.e., $\hat{y}_1^*$, $\hat{y}_2^*$, $\ldots$, $\hat{y}_n^*$) constitute the optimal solution $\hat{x}^*$ to QP (6.1) in view of the conversion and equivalence of QP to the LVI and PLE. The proof is thus completed. □

6.4 Numerical-Experiment Results

To demonstrate the efficacy of the 94LVI algorithm, we implement this algorithm via both MATLAB and C programming languages to solve the general strictly convex QP problem. The numerical experiments are carried out in the MATLAB R2008a environment performed on a personal digital computer equipped with a Pentium(R) Dual-Core E5700 3.00 GHz CPU, 2 GB DDR3 memory, and a Windows 7 Ultimate operating system. The output errors and the computing time of related algorithms are shown in this section.

6.4.1 Efficacy Verification

In this subsection, a numerical example is presented to demonstrate the effectiveness of the 94LVI algorithm for solving QP (6.1) subject to linear equality and bound constraints. That is, we consider the following QP problem:

$$\begin{aligned}
&\text{minimize } 10\hat{x}_1^2+\hat{x}_2^2+\hat{x}_3^2-2\hat{x}_1\hat{x}_2-6\hat{x}_1\hat{x}_3-4\hat{x}_1,\\
&\text{subject to } \hat{x}_1+\hat{x}_2-2\hat{x}_3=0,\\
&\qquad -3\le\hat{x}_1,\hat{x}_2,\hat{x}_3\le 3.
\end{aligned}\tag{6.13}$$

QP (6.13) can be rewritten in a compact matrix-vector form as (6.1), and thus we have the following coefficient matrices/vectors:

$$Q=\begin{bmatrix}20&-2&-6\\-2&2&0\\-6&0&2\end{bmatrix},\quad \hat{q}=\begin{bmatrix}-4\\0\\0\end{bmatrix},\quad J=\begin{bmatrix}1\\1\\-2\end{bmatrix}^{\mathrm{T}},$$

$$\hat{d}=0,\quad \xi^+=\begin{bmatrix}3\\3\\3\end{bmatrix},\quad \xi^-=\begin{bmatrix}-3\\-3\\-3\end{bmatrix}.$$

By employing the 94LVI algorithm program with the PAV being 10^{-6} to solve this QP problem, numerical-experiment results can be summarized below.

- Figure 6.1 shows the solution trajectories of the 94LVI algorithm for solving QP (6.13) with initial state $\hat{y}^0=[(\hat{x}^0)^{\mathrm{T}},(u^0)^{\mathrm{T}}]^{\mathrm{T}}$ being $[0,0,0,0]^{\mathrm{T}}$. As seen from Fig. 6.1, through 277 iterations, the state vector $\hat{y}=[\hat{x}^{\mathrm{T}},u^{\mathrm{T}}]^{\mathrm{T}}$ can converge to the optimal solution $\hat{y}^*=[5/8,9/8,7/8,1]^{\mathrm{T}}$, of which the first n elements of $\hat{y}^*$ constitute the QP's optimal solution $\hat{x}^*=[5/8,9/8,7/8]^{\mathrm{T}}$. Figure 6.2 shows the trajectory of the solution-error $\|\hat{x}^\kappa-\hat{x}^*\|_2$ ($\kappa=0,1,2,\ldots$), which converges to zero within 277 iterations. Note that the solution-error at the final iteration [when employing the 94LVI algorithm to solve QP (6.13)] is about 6.694457×10^{-7} [i.e., less than the PAV 10^{-6}], which validates the accuracy of the presented 94LVI algorithm for solving the QP problem.

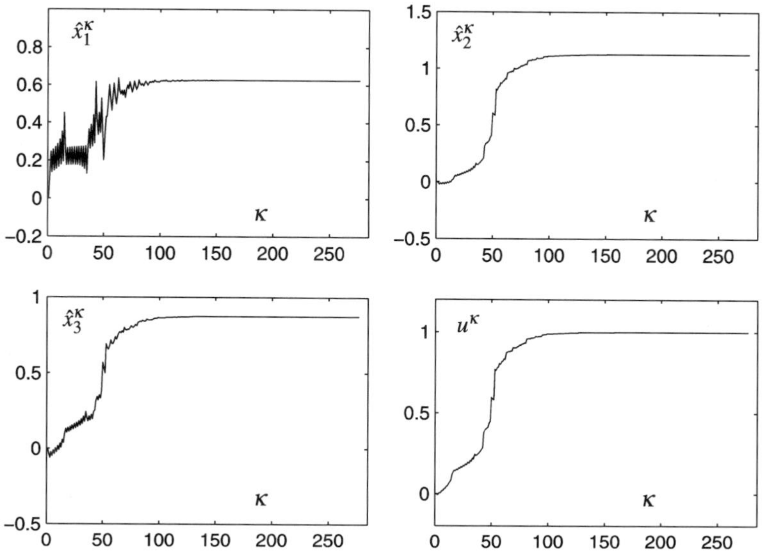

Fig. 6.1 The 94LVI solution trajectories of QP (6.13) starting from initial state $\hat{y}^0 = [0, 0, 0, 0]^T$

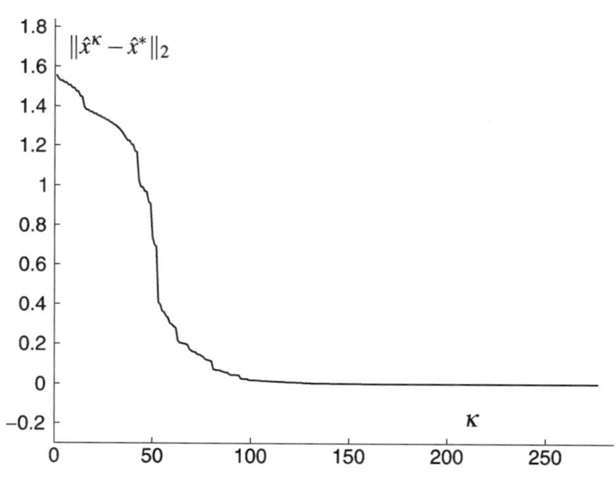

Fig. 6.2 The 94LVI solution-error trajectory of QP (6.13) starting with initial state $\hat{y}^0 = [0, 0, 0, 0]^T$

- For comparison and illustration, the numerical-experiment results with different initial states $\hat{y}^0 = [2, 1/2, -2, 1]^T$ are presented in Figs. 6.3 and 6.4. From Fig. 6.3 we can see that, through 266 iterations, the state vector $\hat{y} = [\hat{x}^T, u^T]^T$ converges to the optimal solution $\hat{y}^* = [5/8, 9/8, 7/8, 1]^T$, and thus $\hat{x}^* = [5/8, 9/8, 7/8]^T$ is the optimal solution of QP (6.13). Figure 6.4 shows the trajectory of the corresponding solution-error $\|\hat{x}^\kappa - \hat{x}^*\|_2$, which converges to zero within 266 iterations. Note that the solution-error at the final iteration is about 6.665722×10^{-7}. The above numerical-experiment results further demonstrate the efficacy and accuracy of the 94LVI algorithm for solving the QP problem.

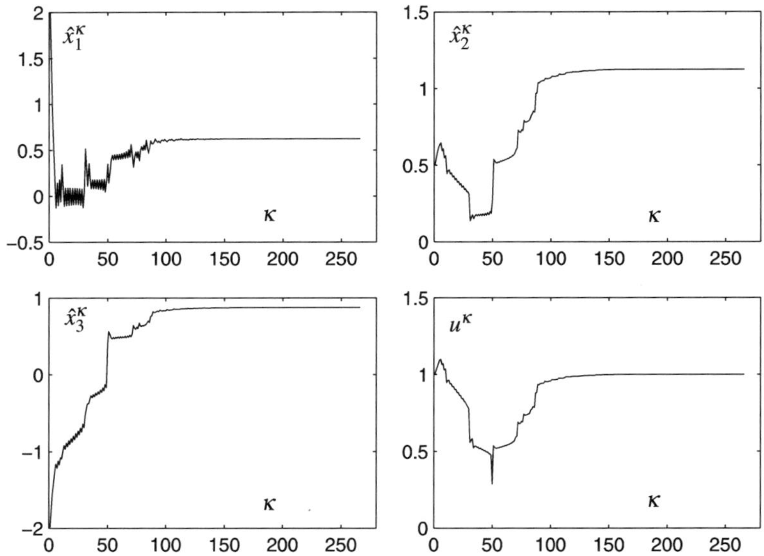

Fig. 6.3 The 94LVI solution trajectories of QP (6.13) from the initial state $\hat{y}^0 = [2, 1/2, -2, 1]^T$

Fig. 6.4 The 94LVI solution-error trajectory of QP (6.13) with initial state $y^0 = [2, 1/2, -2, 1]^T$

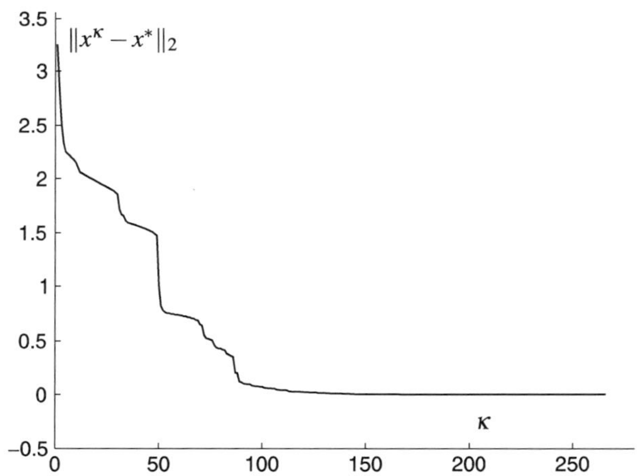

In summary, the above illustrative example has substantiated well the efficacy and accuracy of the 94LVI algorithm for solving such a QP problem.

6.4.2 Comparative Experiments

In this subsection, with coefficients randomly generated, a number of QP problems in the form of QP (6.1) are solved via the active-set algorithm and the 94LVI algorithm. Moreover, both MATLAB- and C-version programs of the 94LVI algorithm

Table 6.1 The running time of the active-set algorithm and 94LVI algorithm and the solution-errors of 94LVI programs (i.e., 94LVIM and 94LVIC programs)

	τ_{AS} (s)	τ_{94LVIM} (s)	$\varepsilon_{\mathrm{94LVIM}}$	τ_{94LVIC} (s)	$\varepsilon_{\mathrm{94LVIC}}$
1	0.715455	0.028507	9.599926×10^{-4}	0.003365	9.513458×10^{-4}
2	0.614927	0.030423	9.606960×10^{-4}	0.003456	9.917611×10^{-4}
3	0.616497	0.023424	9.463763×10^{-4}	0.003546	9.068035×10^{-4}
4	0.356106	0.016992	9.869118×10^{-4}	0.002908	9.869118×10^{-4}
5	0.600025	0.019868	8.605755×10^{-4}	0.003250	9.904625×10^{-4}
6	0.607770	0.023295	7.155840×10^{-4}	0.003017	4.518490×10^{-4}
7	0.616976	0.029874	9.936125×10^{-4}	0.003408	9.936125×10^{-4}
8	0.604829	0.029088	9.850526×10^{-4}	0.004373	9.656885×10^{-4}
9	0.614515	0.019124	6.942093×10^{-4}	0.003064	9.810901×10^{-4}
10	0.600338	0.026207	9.925356×10^{-4}	0.003029	9.962816×10^{-4}
average	0.594744	0.024680	9.095546×10^{-4}	0.003342	9.215806×10^{-4}

are developed and employed to solve such QPs. The computation time τ_{AS} of the active-set algorithm, as well as the computation times τ_{94LVIM}, τ_{94LVIC} and output errors $\varepsilon_{\mathrm{94LVIM}}$, $\varepsilon_{\mathrm{94LVIC}}$ of the 94LVI algorithm per problem, are recorded. Here, ε is defined as $\|e(\hat{y}^{\kappa})\|_2$ [i.e., (6.7)] of the final iteration, while the subscripts $_{\mathrm{AS}}$, $_{\mathrm{94LVIM}}$, and $_{\mathrm{94LVIC}}$ denote the active-set algorithm and the 94LVI algorithm implemented in MATLAB and C, respectively.

First, we show 10 comparison results synthesized by the active-set algorithm and the MATLAB-version 94LVI algorithm in Table 6.1. The values of the coefficient matrices/vectors Q, $\hat{q}$, J, $\hat{d}$, ξ^-, and ξ^+ of the 10 QP problems are all randomly generated through MATLAB function "rand()". In this case, we set $n = 3$ and $m = 1$, with the PAV being 10^{-3}. As seen from Table 6.1, for such 10 randomly generated QP problems, compared with the conventional active-set algorithm, by adopting the MATLAB-version 94LVI algorithm, the running time is shortened by about 20 times. In addition, all errors of final solutions are less than the PAV.

Second, we also show 10 numerical-experiment results using C program in Table 6.1. The coefficient matrices/vectors of these 10 QP problems are set the same as above. As seen from Table 6.1, the runtime of the C-version 94LVI algorithm is about 8 times in average shorter than the MATLAB-version one. In addition, all the errors of the solutions synthesized by the C-version 94LVI algorithm are less than the PAV (i.e., 10^{-3}) again.

In summary, the above numerical-experiment results have sufficiently substantiated the efficacy and accuracy of the 94LVI algorithm for solving such QP problems. Compared to the active-set algorithm, the 94LVI algorithm can obtain better performance.

6.5 Chapter Conclusions

In this chapter, a numerical quadratic-programming algorithm (i.e., the 94LVI algorithm) has been presented and investigated to solve the strictly convex quadratic-programming problem subject to linear equality and bound constraints. The global convergence of the 94LVI algorithm has been proved. Moreover, the numerical-experiment results have demonstrated the efficacy, accuracy, and superiority of the presented 94LVI algorithm for QP solving. To further demonstrate the advantages, 10 more comparison results between the active-set algorithm and the 94LVI algorithm have been provided, illustrating the higher accuracy and faster convergence of the 94LVI algorithm.

References

1. Johansen TA, Fossen TI, Berge SP (2004) Constrained nonlinear control allocation with singularity avoidance using sequential quadratic programming. IEEE Trans Control Syst Technol 12(1):211–216
2. Grudinin N (1998) Reactive power optimization using successive quadratic programming method. IEEE Trans Power Syst 13(4):1219–1225
3. Wang J, Zhang Y (2004) Recurrent neural networks for real-time computation of inverse kinematics of redundant manipulators. In: Sincak P, Vascak J, Hirota K (eds) Machine intelligence: quo vadis? World Scientific, Singapore, pp 299–319
4. Leithead WE, Zhang YN (2007) O(N^2)-operation approximation of covariance matrix inverse in Gaussian process regression based on quasi-Newton BFGS method. Commun Stat, Simul Comput 36(2):367–380
5. Zhang Y (2006) Towards piecewise-linear primal neural networks for optimization and redundant robotics. In: Proceedings of IEEE international conference on networking, sensing and control, pp 374–379
6. Zhang Y, Wang J (2002) A dual neural network for constrained torque optimization of kinematically redundant manipulators. IEEE Trans Syst Man Cybern, Part B, Cybern 32(5):654–662
7. Zhang Y, Wang J (2004) Obstacle avoidance of kinematically redundant manipulators using a dual neural network. IEEE Trans Syst Man Cybern, Part B, Cybern 34(1):752–759
8. Powell MJD (1969) A method for nonlinear constraints in minimization problems. In: Fletcher R (ed) Optimization. Academic Press, London
9. Hestens MR (1969) Multiplier and gradient methods. J Optim Theory Appl 4(5):303–320
10. Fletcher R (1971) A general quadratic programming algorithm. IMA J Appl Math 7(1):76–91
11. Jin Z, Bai Y, Han B (2010) A weighted-path-following interior-point algorithm for convex quadratic optimization. OR Trans 14(1):55–65
12. Zhao S, Fei P, Li J (2001) A projection and contraction method for convex quadratic programming. J Wuhan Univ Technol 47(1):22–24
13. Zhang Y, Li X, Zhang Z, Li J (2012) An LVI-based numerical algorithm for solving quadratic programming problems. OR Trans 16(1):21–30
14. He B (1994) A new method for a class of linear variational inequalities. Math Program 66(1–3):137–144
15. Zhang Y (2005) On the LVI-based primal–dual neural network for solving online linear and quadratic programming problems. In: Proceedings of American control conference, pp 1351–1356

16. Zhang Y, Cai B, Zhang L, Li K (2008) Bi-criteria velocity minimization of robot manipulators using a linear variational inequalities-based primal–dual neural network and PUMA560 example. Adv Robot 22(13–14):1479–1496
17. He B (1994) Solving a class of linear projection equations. Numer Math 68(1):71–80
18. He B (1992) A projection and contraction method for a class of linear complementarity problem and its application in convex quadratic programming. Appl Math Optim 25(3):247–262

Chapter 7
Numerical Algorithm E47

Abstract In this chapter, based on the linear variational inequality (LVI), a numerical algorithm (termed E47 algorithm), which won the ICAL2011 best paper award (Zhang et al. in Proceedings of IEEE International Conference on Automation and Logistics, pp. 125–130, 2011), is presented and investigated to solve the QP problem that is simultaneously subject to linear equality, inequality, and bound constraints. Note that the constrained QP problem can be converted to a linear variational inequality and then to a piecewise-linear equation (PLE). The E47 algorithm is then adapted to solving the resultant PLE, and thus the optimal numerical solution to the QP problem is obtained. In addition, the global linear convergence of such an E47 algorithm is proved. The numerical comparison results between such an E47 algorithm and the active-set algorithm are further provided. The efficacy and superiority of the presented E47 algorithm for QP solving are substantiated.

7.1 Introduction

In numerous areas of applications, such as engineering and scientific research, manufacturing, and economic statistics, QP problems are widely encountered and investigated by researchers; see, e.g., [2–6]. Especially in recent years, different kinds of online solution methods aiming at QP problems have been proposed and developed, e.g., the recurrent neural network (RNN) and the numerical algorithm (NA) [3, 7–10]. In addition to the neural networks shown in Chaps. 4 and 5, in [7], two new classes of high-performance networks with economical analog multipliers are proposed for solving the QP problems. In [8] as well as in Sect. 5.3, a simplified LVI-based primal–dual neural network (S-LVI-PDNN) is presented as a powerful tool for solving online the QP problems.

However, in the literature (e.g., [10, 11] and most textbooks), people often discuss the QP problems subject to one or two kinds of typical constraints (e.g., only the equality constraint and/or the inequality constraint). In actual manufacturing and economic areas, the QP problems are often simultaneously subject to all (or to say, three) kinds of constraints [12], i.e., the equality constraint, the inequality constraint, and the bound constraint. It would observably increase the computational complexity if we simply extend and apply the existing methods (which are designed for solving the QP problems subject to one or two kinds of constraints) to solving the

Y. Zhang, Z. Zhang, *Repetitive Motion Planning and Control of Redundant Robot Manipulators*, DOI 10.1007/978-3-642-37518-7_7,

general QP problems with three kinds of constraints [8]. Thus, the existing methods may not be efficient enough in solving the general QP problems subject to three kinds of constraints and may lose the practical value. Motivated by engineering applications of the QP, e.g., in robotics [3, 13], we still discuss the following general problem formulation in this chapter:

$$\text{minimize} \quad \hat{x}^{\mathrm{T}} Q \hat{x}/2 + \hat{q}^{\mathrm{T}} \hat{x}, \tag{7.1}$$

$$\text{subject to} \quad J\hat{x} = \hat{d}, \tag{7.2}$$

$$A\hat{x} \leq \hat{b}, \tag{7.3}$$

$$\xi^{-} \leq \hat{x} \leq \xi^{+}, \tag{7.4}$$

where $\hat{x} \in R^n$ is the decision vector to be obtained, and $Q \in R^{n\times n}$ is assumed to be a positive definite symmetric matrix. The coefficients are defined respectively as $\hat{q} \in R^n$, $J \in R^{m\times n}$, $\hat{d} \in R^m$, $A \in R^{\dim(\hat{b})\times n}$, and $\hat{b} \in R^{\dim(\hat{b})}$, with $\dim(\hat{b})$ denoting the dimension of a vector $\hat{b}$. Besides, ξ^- and ξ^+ denote the lower and upper bounds of $\hat{x}$, respectively.

Before proceeding, let us review again the existing QP solvers in relatively more detail, especially in the authors' background of RNN. In the existing literature, most QP solvers can be classified to be either parallel-processing methods such as neural networks or serial-processing numerical algorithms [8, 11, 12]. Some of existing neural networks contain penalty parameters, of which the stable equilibrium points (corresponding to the solutions of optimization problems) can be achieved only when the penalty parameters are infinitely large. This is almost impossible when we solve the problem numerically [7]. Reference [11] proposes a low-order discrete-time recurrent neural network for solving high-dimension QP problems, especially for the case where the number of decision variables is close to the number of constraints. However, the QP problem depicted in [11] only involves the equality constraint. Some numerical algorithms are also proposed to solve the QP problem, e.g., the active-set algorithm [14], the quasi-Newton method [15], and the Lagrangian algorithm [16]. Based on the Karush–Kuhn–Tacker optimality conditions and the projection operator [8], researchers have improved the traditional dual neural network (DNN) [12, 13]. However, the improved DNN still has the disadvantage that, in the process of solving the QP problems, the inversion of coefficient matrix Q is required, and thus it can only handle strictly convex QP problems with fixed and/or easy-to-invert coefficient matrix Q [8, 12, 13].

In order to solve general QP problems (subject to equality, inequality, and bound constraints, simultaneously) and obtain optimal solutions for robot control, a numerical algorithm (termed E47 algorithm) based on the linear variational inequality (LVI) is developed, presented, and investigated in this chapter [1]. Note that, for simplicity, we call it E47 algorithm, which is named after Eqs. (4)–(7) of [17] [which was designed originally to solve linear projection equations]. To do so, the general QP problems are first converted via the important "bridge" theorem into linear variational inequalities (LVI) [8, 18]. The E47 algorithm is then adapted to solving the resultant piecewise-linear equations, the LVI and QP problems. Furthermore, the numerical comparison results between such an E47 algorithm and the conventional active-set algorithm are provided. The conducted numerical experiments demonstrate well the efficacy and superiority of such an E47 algorithm for QP solving.

The remainder of this chapter is organized into four sections. The important "bridge" theorem and the E47 algorithm are presented in Sect. 7.2. Theoretical results and a simple proof of the E47 algorithm are presented in Sect. 7.3. The global convergence behaviors are further investigated via numerical experiments in Sect. 7.4. Section 7.5 concludes this chapter with final remarks. Before ending this section, we list the main novelties and contributions of the work as follows.

1. This chapter considers the QP problems subject to all kinds of linear constraints (i.e., equality, inequality, and bound constraints) simultaneously. This is quite different from the conventional situation of handling one or two kinds of constraints in most literature and textbooks.
2. An LVI-based numerical method with a very simple and implementable structure is proposed to solve accurately the QP problems. This numerical algorithm can be viewed as an efficient discrete-time model of LVI-PDNN, which takes researchers (including the authors) decades to finally find. Note that this interdisciplinary LVI-based numerical method is very novel and different from the results of pure RNN or NA research community.
3. The core equations (4)–(7) of [17] were originally designed for solving linear projection equations, but now they are adapted to solving QP problems. That is, the authors of this chapter have discovered the new area and utility of such core equations.
4. The global linear convergence of the E47 algorithm is presented for QP solving, with a simple proof given by the authors' engineering-type understanding.
5. Numerical experiments substantiate well the efficacy and superiority of the E47 algorithm for QP solving (it is worth mentioning that, on average, the E47 algorithm implemented in C program is 168 times faster than MATLAB built-in function "Quadprog" and that the former is thus now applied in real-time robot control).

7.2 E47 Algorithm for QP Solving

With the aid of dual decision variables, for the primal QP problem (5.7)–(5.10), its dual QP problem can be derived via duality theory [3, 8, 18]. For each constraint, such as (5.8), (5.9), and (5.10), the dual decision variables can be defined as the Lagrangian multipliers. In addition, for bound constraint (5.10), we use an elegant treatment to cancel the dual decision variables to reduce the QP-solver complexity. Thus, we have the "bridge" theorem [i.e., the conversion from QP (5.7)–(5.10) to linear variational inequality (LVI) (5.11)].

Solving LVI (5.11) is equivalent to finding a zero point of the following error function [which is actually the negation of the left-hand side of PLE (5.14)]:

$e(\hat{y}) := \hat{y} - \mathscr{P}_\Omega(\hat{y} - (H\hat{y} + \hat{p}))$. The following numerical algorithm (i.e., E47 algorithm) is then employed to solve PLE (5.14), LVI (5.11), and QP (5.7)–(5.10) as follows. Let the optimal solution set $\Omega^* = \{\hat{y}^* | \hat{y}^* \text{ is a solution of (5.14)}\}$. Given $\hat{y}^0 \in R^{n+m+\dim(\hat{b})}$, for iteration indices $\kappa = 0, 1, 2, 3, \ldots$, if $\hat{y}^\kappa \notin \Omega^*$, then we have the following E47 iteration formula:

$$\hat{y}^{\kappa+1} = \mathscr{P}_\Omega\big(\hat{y}^\kappa - \hat{\rho}(\hat{y}^\kappa) g(\hat{y}^\kappa)\big), \tag{7.5}$$

$$\text{with} \quad g(\hat{y}^\kappa) := H^{\mathrm{T}} e(\hat{y}^\kappa) + H\hat{y}^\kappa + \hat{p}, \tag{7.6}$$

$$\hat{\rho}(\hat{y}^\kappa) := \frac{\|e(\hat{y}^\kappa)\|_2^2}{\|(H^{\mathrm{T}} + I)e(\hat{y}^\kappa)\|_2^2}. \tag{7.7}$$

The main advantage of the E47 algorithm is that it possesses a very simple and implementable structure, in addition to the ability to handle the piecewise-linear equation (5.14) and QP (5.7)–(5.10) (which is subject to all kinds of linear constraints); whereas some other algorithms (e.g., [7, 9]) may only solve special cases of (5.14). Evidently, each iteration of the E47 algorithm consists essentially of only two projections to set Ω and only two matrix–vector products [i.e., $H^{\mathrm{T}} e(\hat{y})$ and $H\hat{y}$]. Therefore, the E47 algorithm, generally of $O((n + m + \dim(\hat{b}))^2)$ operations, could also allow the optimal exploitation of the sparsity of matrix H and may thus be efficient for very large sparse problems.

7.3 Theoretical Results and Simple Proof

In this section, by following [17, 19], the global linear convergence of E47 algorithm (7.5)–(7.7) is proved as below.

Theorem 7.1 *For all* $\hat{y}^* \in \Omega^*$, *the sequence* $\{\hat{y}^\kappa\}$ (*with iteration indices* $\kappa = 0, 1, 2, 3, \ldots$ *used as the superscript*) *generated by the* E47 *algorithm* (7.5)–(7.7) *satisfies*

$$\|\hat{y}^{\kappa+1} - \hat{y}^*\|_2^2 \le \|\hat{y}^\kappa - \hat{y}^*\|_2^2 - \hat{\rho}(\hat{y}^\kappa)\|e(\hat{y}^\kappa)\|_2^2. \tag{7.8}$$

Proof See [19] and can be generalized from it. □

Theorem 7.2 *The sequence* $\{\hat{y}^\kappa\}$ *generated by the* E47 *algorithm* (7.5)–(7.7) *converges to an optimal solution* $\hat{y}^*$.

Proof According to (7.7) and Chap. 6, we have $\hat{\rho}(\hat{y}) \geqslant 1/\|H^{\mathrm{T}} + I\|_{\mathrm{F}}^2 = \delta > 0$. Then, from Theorem 7.1 [i.e., (7.8)] we get that, for all $\hat{y}^* \in \Omega^*$,

$$\|\hat{y}^{\kappa+1} - \hat{y}^*\|_2^2 \le \|\hat{y}^\kappa - \hat{y}^*\|_2^2 - \delta\|e(\hat{y}^\kappa)\|_2^2, \tag{7.9}$$

which implies that, if $\|e(\hat{y}^\kappa)\|_2 = 0$, then $\hat{y}^\kappa = \hat{y}^*$, $\|\hat{y}^{\kappa+1} - \hat{y}^*\|_2^2 \leq \|\hat{y}^\kappa - \hat{y}^*\|_2^2$, and $y^{\kappa+1} = \hat{y}^\kappa = \hat{y}^*$, and that, if $\|e(\hat{y}^\kappa)\|_2 \neq 0$, then $\|\hat{y}^{\kappa+1} - \hat{y}^*\|_2^2 < \|y^\kappa - \hat{y}^*\|_2^2 < \cdots < \|\hat{y}^0 - \hat{y}^*\|_2^2$. So, in the situation of $\|e(\hat{y}^\kappa)\|_2 \neq 0$, summing up (7.8) from $\kappa = 0$ to infinity, we get $\delta \sum_{\kappa=0}^{\infty} \|e(y^\kappa)\|_2^2 \leq \|\hat{y}^0 - \hat{y}^*\|_2^2 < \infty$, which means $\lim_{\kappa\to\infty} e(\hat{y}^\kappa) = 0$ and $\lim_{\kappa\to\infty} \hat{y}^\kappa = \hat{y}^*$. The proof is thus complete. □

7.4 Numerical-Experiment Results

To demonstrate the efficacy of E47 algorithm (7.5)–(7.7), we implement this algorithm via both MATLAB and C program languages to solve general QP problems. The numerical experiments are carried out in MATLAB R2008a environment performed on a personal digital computer equipped with an INTEL Core(TM) Duo E4500 2.20 GHz CPU, 2 GB DDR3 memory, and the Windows XP Professional operating system. The final output errors and the computing time of algorithms are shown in this section.

7.4.1 *QP Problem with Specified Coefficients*

In this subsection, without loss of generality, the following QP problem is considered with corresponding coefficient parameters specified:

$$\text{minimize} \quad \hat{x}^{\mathrm{T}} Q \hat{x}/2 + \hat{q}^{\mathrm{T}} \hat{x}, \tag{7.10}$$

$$\text{subject to} \quad J\hat{x} = \hat{d}, \tag{7.11}$$

$$A\hat{x} \leq \hat{b}, \tag{7.12}$$

$$\xi^- \leq \hat{x} \leq \xi^+, \tag{7.13}$$

where

$$Q = \begin{bmatrix} 22 & -2 & 6 \\ -2 & 2 & 0 \\ 6 & 0 & 2 \end{bmatrix}, \quad \hat{q} = \begin{bmatrix} -4 \\ 0 \\ 0 \end{bmatrix}, \quad J = \begin{bmatrix} 2 \\ 2 \\ 1 \end{bmatrix}^{\mathrm{T}}, \quad \hat{d} = 0,$$

$$A = \begin{bmatrix} -1 & 1 & 0 \\ 3 & 0 & 1 \end{bmatrix}, \quad \hat{b} = \begin{bmatrix} -1 \\ 4 \end{bmatrix}, \quad -\xi^- = \xi^+ = \begin{bmatrix} 6 \\ 6 \\ 6 \end{bmatrix}.$$

Throughout this chapter, the prescribed error criterion $\|e(\hat{y}^\kappa)\|_2 \leq 10^{-3}$ is used, which is accurate for most actual applications (for example, in robotics, 10^{-3} in meters means that the high precision of less than 1 mm is achieved). To show the intermediate results and the convergence performance of the E47 algorithm (7.5)–(7.7), the value of $\hat{x}^\kappa$ (which is a column vector made of the first n elements of $\hat{y}^\kappa$)

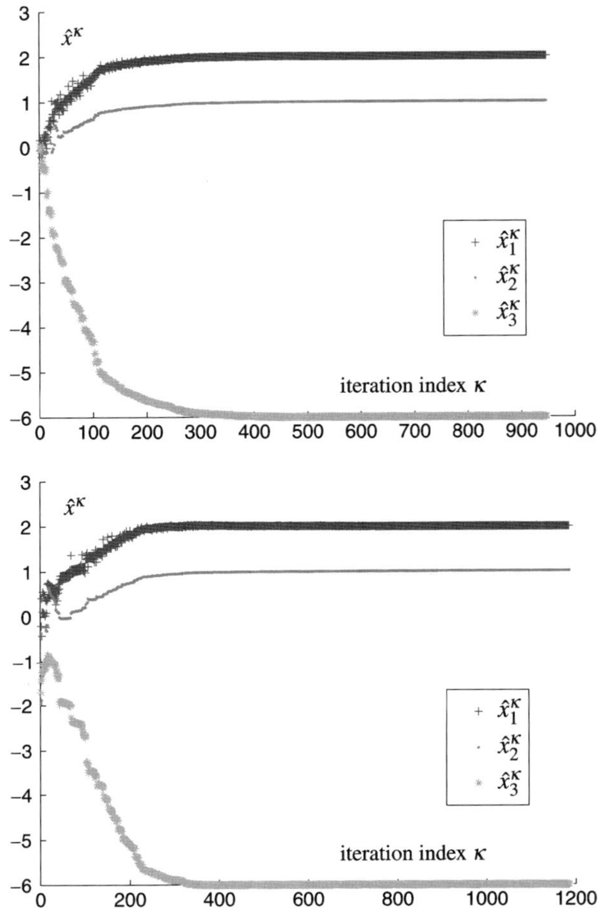

Fig. 7.1 Solving QP problem (7.10)–(7.13) by the presented E47 algorithm (7.5)–(7.7) with two different initial states, i.e., initial vectors $\hat{y}^0 = (0, 0, 0, 0, 0, 0)^{\mathrm{T}}$ and $\hat{y}^0 = (-2, -2, -2, 2, 2, 2)^{\mathrm{T}}$

is tracked and depicted in Fig. 7.1. Specifically, two numerical experiments with different initial states are conducted to demonstrate the efficacy of this E47 algorithm, i.e., shown in the two graphs of Fig. 7.1. As seen from the figure, starting from two different initial states, the E47 algorithm both converges to the same unique optimal solution. In addition, we see that it needs about 500 iterations to achieve the optimal solution of the QP problem (7.10)–(7.13) with high accuracy. Note that the optimal solution of the QP problem (7.10)–(7.13) is $\hat{x}^* = [2, 1, -6]^{\mathrm{T}}$. Figure 7.2 further illustrates the corresponding computational error, i.e., $\|\hat{x}^\kappa - \hat{x}^*\|_2$ over κ. From the figure we see that $\|\hat{x}^\kappa - \hat{x}^*\|_2$ converges to 0 as the iteration index κ increases. Thus, we can say that the E47 iteration process is convergent, and the E47 algorithm is successful in solving the QP problem (7.10)–(7.13).

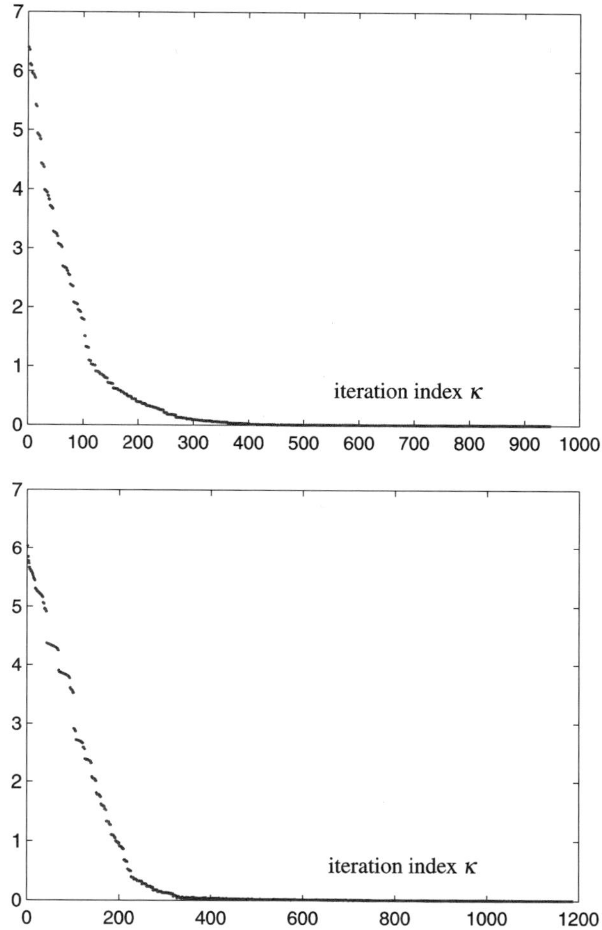

Fig. 7.2 Computational error $\|\hat{x}^{\kappa} - \hat{x}^*\|_2$ of the E47 algorithm (7.5)–(7.7) solving QP problem (7.10)–(7.13) with two different initial states, i.e., initial vectors $\hat{y}^0 = (0, 0, 0, 0, 0, 0)^{\mathrm{T}}$ and $\hat{y}^0 = (-2, -2, -2, 2, 2, 2)^{\mathrm{T}}$

7.4.2 QP Problems with Randomly Generated Coefficients

Following the above numerical experiments with coefficients specified, we further investigate the E47 algorithm for solving the QP problems with randomly generated coefficients. That is to say, the values of coefficients Q, $\hat{q}$, J, $\hat{d}$, A, $\hat{b}$, ξ^-, and ξ^+ are all randomly generated by using MATLAB function "rand()". The prescribed error criterion $\|e(\hat{y}^{\kappa})\|_2 \leq 10^{-3}$ is used again, being the same as before. For comparison and illustration, we not only develop the MATLAB program for the E47 algorithm, but also implement the algorithm in C program. For these randomly generated solvable quadratic-programming problems, the presented E47 algorithm in this chapter shows great advantages. That is, we conduct numerous numerical experiments of E47 algorithm (7.5)–(7.7) using both MATLAB and C programs, and obtain the desired results also in this random situation. The results about the average computing time and the final output errors of 20 experiments are shown

Table 7.1 The computing time (in seconds) and the final output error of the E47 algorithm for solving randomly generated solvable QP problems

Test #	$\text{Time}_{\text{MATLAB}}$	Time_{C}	$\text{Error}_{\text{MATLAB}}$ (i.e., $\varepsilon_{\text{E47M}}$)	Error_{C} (i.e., $\varepsilon_{\text{E47C}}$)
1	0.02181373102	0.00394436004	$8.90078534207947 \times 10^{-4}$	$8.90078534207018 \times 10^{-4}$
2	0.01721218900	0.00348286049	$7.33399346422728 \times 10^{-4}$	$7.33399346422764 \times 10^{-4}$
3	0.02658747700	0.00388566885	$8.75932325689998 \times 10^{-4}$	$8.75932512358256 \times 10^{-4}$
4	0.02908979002	0.00399562775	$7.40338441679319 \times 10^{-4}$	$9.63347628258944 \times 10^{-4}$
5	0.02504048858	0.00392329257	$9.10313617957184 \times 10^{-4}$	$9.10313637125233 \times 10^{-4}$
6	0.07280702067	0.00652331398	$9.74053797216939 \times 10^{-4}$	$9.79271354792625 \times 10^{-4}$
7	0.01841975193	0.00359698476	$7.57917212034314 \times 10^{-4}$	$7.57917215389154 \times 10^{-4}$
8	0.05837775869	0.00513098081	$9.65815016827337 \times 10^{-4}$	$9.53240506576691 \times 10^{-4}$
9	0.02287355595	0.00379865711	$6.86243744747358 \times 10^{-4}$	$8.91371467172357 \times 10^{-4}$
10	0.01800682837	0.00353762189	$9.83050001295021 \times 10^{-4}$	$9.83050001295021 \times 10^{-4}$
11	0.02810104715	0.00474342544	$9.70253520703793 \times 10^{-4}$	$9.70253520694082 \times 10^{-4}$
12	0.01795118477	0.00387432559	$9.47721317780562 \times 10^{-4}$	$9.47721317814435 \times 10^{-4}$
13	0.03199583371	0.00422726974	$9.60886027586627 \times 10^{-4}$	$9.38488935465161 \times 10^{-4}$
14	0.04852938228	0.00517810318	$9.25414059359843 \times 10^{-4}$	$9.58079766524444 \times 10^{-4}$
15	0.01848792173	0.00386589960	$7.65807228108965 \times 10^{-4}$	$7.65807228108965 \times 10^{-4}$
16	0.02710185325	0.00394153801	$7.72849859123260 \times 10^{-4}$	$9.12385893963973 \times 10^{-4}$
17	0.01733093478	0.00315919427	$8.18845196308935 \times 10^{-4}$	$8.18845196308536 \times 10^{-4}$
18	0.15382974090	0.00911291945	$9.78920394368993 \times 10^{-4}$	$9.74877116643424 \times 10^{-4}$
19	0.03304986922	0.00516568726	$9.61606698146433 \times 10^{-4}$	$9.61606698152893 \times 10^{-4}$
20	0.02073031734	0.00363290423	$7.89740536959277 \times 10^{-4}$	$7.89740948801000 \times 10^{-4}$
Average	0.03536683382	0.00443603175	$8.70459343826242 \times 10^{-4}$	$8.98786441303749 \times 10^{-4}$

Table 7.2 Comparison on the computing time (in seconds) and the solution-difference between E47 and active-set algorithms

Test #	Time_{AS} (i.e., τ_{AS})	Time_{E47} (i.e., τ_{E47M})	Solution-difference
1	0.876790946539562	0.028101047153820	$5.366265848423164\times10^{-4}$
2	0.860737523923227	0.017951184769335	$2.242231377880266\times10^{-4}$
3	0.863932241674717	0.031995833713067	$3.524569856262061\times10^{-4}$
4	0.867815193755354	0.048529382279496	$4.108643061441598\times10^{-4}$
5	0.873233059165558	0.018487921732315	$1.775320795459161\times10^{-4}$
6	0.857095121028744	0.027101853252648	$2.450439820686725\times10^{-4}$
7	0.860923797914806	0.017330934783005	$1.731857408659998\times10^{-4}$
8	0.565073070648707	0.153829740900060	$0.001596302347334\times10^{-0}$
9	0.870159758853872	0.033049869219693	$3.979301303627689\times10^{-4}$
10	0.859110024971292	0.020730317335909	$2.396982274348591\times10^{-4}$
Average	0.835487073847584	0.039710808513935	$4.353863522012926\times10^{-4}$

in Table 7.1, wherein $\text{Time}_{\text{MATLAB}}$ and Time_{C} can also be denoted respectively by τ_{E47M} and τ_{E47C} in the notation of Sect. 6.4 of Chap. 6. As seen from the table, the computing time of the E47 algorithm is relatively very small (i.e., with the average one of the C program being 0.0044 s, in other words, 4.4 milliseconds). In addition, the final output errors are tiny (less than 10^{-3}). Another observation from the table of these random numerical experiments is that the E47 algorithm implemented in the C program is usually much (around 8 times) faster than that implemented in the MATLAB program (i.e., via the M-file). In summary, the fast solution speed and the tiny computational errors substantiate the efficacy of the E47 algorithm presented in this chapter.

7.4.3 Comparison Between E47 and Active-Set Algorithms

To better illustrate the advantages of the presented E47 algorithm, a series of general QP problems are solved and presented in this subsection via both E47 algorithm (7.5)–(7.7) and the active-set algorithm. The computing time and the solution-difference for solving the randomly generated QP problems are recorded for both E47 and active-set algorithms. In addition, the comparison on the average computing time and the average solution-difference between the E47 algorithm and the active-set algorithm is presented in Table 7.2. Note that the active-set algorithm adopted and tested here is actually a built-in function of MATLAB under the name of "Quadprog". The 10 groups of comparison tests in the table (i.e., Table 7.2) actually correspond to Tests 11 through 20 in Table 7.1, wherein, without loss of generality, $n = 3$, $m = 1$, and $\dim(\hat{b}) = 2$.

As seen from Table 7.2, the average computing time for solving a QP problem by the E47 algorithm (MATLAB code) is about 0.04 s, while the average computing time of the active-set algorithm is about 0.84 s. These results demonstrate that the computing time of the E47 algorithm (MATLAB code) is about 21 times faster than that of the active-set algorithm. So, we can estimate that the E47 algorithm implemented in the C program is $21 \times 8 = 168$ times faster than the MATLAB built-in function "Quadprog". In summary, the numerical-comparison results substantiate well the efficacy and superiority of the presented E47 algorithm for QP solving. Compared to the active-set algorithm, the E47 algorithm can achieve almost the same accuracy (i.e., solution effectiveness) in solving the randomly generated solvable QP problems, which is shown in the fourth column of Table 7.2.

7.5 Chapter Conclusions

In this chapter, an LVI-based numerical method (i.e., the E47 algorithm) has been presented to solve the general QP problems subject to linear equality, inequality, and bound constraints. The QP problems with both specified and randomly generated coefficients have been investigated and solved. The numerical-testing results have demonstrated the efficacy and superiority of the presented E47 algorithm for QP solving. In addition, to further demonstrate the advantages, the comparison results between the E47 algorithm and the active-set algorithm have been provided, which illustrate the high accuracy and the (168 times) faster convergence of the E47 algorithm. The numerical results and discussions have provided interesting insights and new answers to related quadratic-programming problems.

References

1. Zhang Y, Fu S, Zhang Z, Xiao L, Li X (2011) On the LVI-based numerical method (E47 algorithm) for solving quadratic programming problems. In: Proceedings of IEEE international conference on automation and logistics, pp 125–130
2. Dwyer T, Koren Y, Marriott K (2006) Drawing directed graphs using quadratic programming. IEEE Trans Vis Comput Graph 12(4):536–548
3. Zhang Y, Ge SS, Lee TH (2004) A unified quadratic programming based dynamical system approach to joint torque optimization of physically constrained redundant manipulators. IEEE Trans Syst Man Cybern, Part B, Cybern 34(5):2126–2132
4. Plumlee JH, Bevly DM, Hodel AS (2004) Control of a ground vehicle using quadratic programming based control allocation techniques. In: Proceedings of American control conference, pp 4704–4709
5. Wang M, Tzeng GH, Jane TD (2009) A fuzzy ARIMA model by using quadratic programming approach for time series data. Int J Inf Syst Logist Manag 5(1):41–51
6. Wright SE, Foley JA, Huges JM (2000) Optimization of site occupancies in minerals using quadratic programming. Am Mineral 85:524–531
7. Malek A, Oskoei HG (2005) Numerical solutions for constrained quadratic problems using high-performance neural networks. Appl Math Comput 169(1):451–471
8. Zhang Y, Li Z, Tan H, Fan Z (2007) On the simplified LVI-based primal-dual neural network for solving LP and QP problems. In: Proceedings of IEEE international conference on control and automation, pp 3129–3134

9. Moraru V (1997) An algorithm for solving quadratic programming problems. Comput Sci J Mold 5(2):223–235
10. Dominguez J, Gonzalez-Lima MD (2006) A primal–dual interior-point algorithm for quadratic programming. Numer Algorithms 42(1):1–30
11. Liao W, Wang J, Wang J (2010) A lower order discrete-time recurrent neural network for solving high order quadratic problems with equality constraints. Lect Notes Comput Sci 6063:193–198
12. Zhang Y, Wang J (2002) A dual neural network for convex quadratic programming subject to linear equality and inequality constraints. Phys Lett A 298:271–278
13. Zhang Y, Wang J (2002) A dual neural network for constrained torque optimization of kinematically redundant manipulators. IEEE Trans Syst Man Cybern, Part B, Cybern 32(5):654–662
14. Byrd RH, Waltz RA (2011) An active-set algorithm for nonlinear programming using parametric linear programming. Optim Methods Softw 26:47–66
15. Moraru V (1995) Quasi-Newton methods for solving nonlinear programming problems. Comput Sci J Mold 3(3):263–277
16. Conn AR, Gould N, Sartenaer A, Toint PL (1996) Convergence properties of an augmented Lagrangian algorithm for optimization with a combination of general equality and linear constraints. SIAM J Optim 6(3):674–703
17. He B (1994) Solving a class of linear projection equations. Numer Math 68:71–80
18. Zhang Y (2005) On the LVI-based primal–dual neural network for solving online linear and quadratic programming problems. In: Proceedings of American control conference, pp 1351–1356
19. He B (1994) A new method for a class of linear variational inequalities. Math Program 66:137–144

Part III
Robot Simulations and Experiments

Chapter 8
Examples of Planar Multilink Manipulators

Abstract In Chaps. 2 and 3, we have theoretically demonstrated the effectiveness of such a physically constrained RMP scheme (and its solvers) on solving the joint-angle drift problem. In this chapter, a dual neural network (introduced in Chap. 4) and an LVI-based primal–dual neural network (introduced in Chap. 5) are presented for online repetitive motion planning (RMP) of redundant robot manipulators (with multilink planar manipulators as examples). As real-time QP solvers, the aforementioned two kinds of neural networks both have piecewise-linear dynamics and can globally exponentially converge to the optimal solution of strictly-convex quadratic programs. Furthermore, the neural-network-based physically constrained RMP scheme is simulated based on the multilink planar robot manipulators. Computer-simulation results substantiate the theoretical analysis and also show the effective remedy of the joint-angle drift problem of robot manipulators.

8.1 Three-Link Planar Robot Manipulator

In this section, the physically constrained RMP scheme is employed to handle the joint-angle drift problem of a redundant three-link planar robot arm when its end-effector moves along different types of paths (e.g., a Lissajous path, a triangular path, and an elliptical path) [1]. The physical constraints such as joint limits and joint-velocity limits are considered as well. The QP solver used in the computer simulations is LVI-PDNN (5.35), and the simulations are performed on a three-link planar robot arm with different types of end-effector paths tracked. The three-link planar robot arm is of three DOF (with one degree of redundancy here) and operates in the two-dimensional (horizontal) plane. The joint physical limits are given as $\theta^{\pm} = [\pm\pi/3, \pm\pi/3, \pm\pi/3]^{\mathrm{T}}$ rad and $\dot{\theta}^{\pm} = [\pm\pi, \pm\pi, \pm\pi]^{\mathrm{T}}$ rad/s. In addition, the initial state $\theta(0) = [\pi/12, \pi/12, \pi/6]^{\mathrm{T}}$ rad. A series of computer simulations conducted demonstrate the effectiveness of such an RMP scheme-formulation and its corresponding neural-network solver on robot motion planning and control.

8.1.1 Lissajous Path Tracking

First, the end-effector of the three-link planar robot arm is expected to track a Lissajous path, with the frequency in the X-axis being duple of that of the Y-axis and

Y. Zhang, Z. Zhang, *Repetitive Motion Planning and Control of Redundant Robot Manipulators*, DOI 10.1007/978-3-642-37518-7_8,

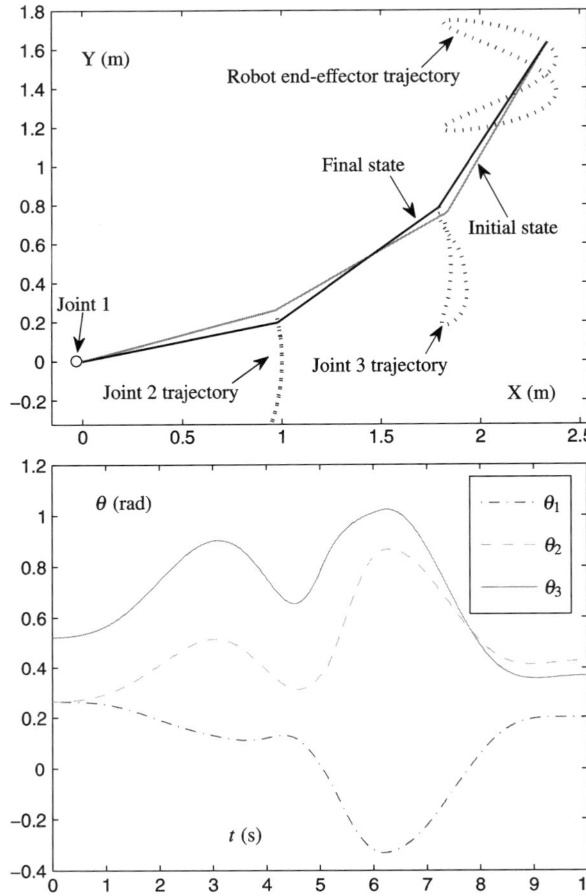

Fig. 8.1 Nonrepetitive motion trajectories of three-link planar robot arm when its end-effector tracking a Lissajous path without considering the drift-free criterion [i.e., $\lambda = 0$ in (2.2)]

Table 8.1 Joint displacement of three-link robot arm when its end-effector tracking a Lissajous path without considering the drift-free criterion [i.e., $\lambda = 0$ in (2.2)]

Joint	$\theta(0)$ (rad)	$\theta(10)$ (rad)	$\theta(10) - \theta(0)$
θ_1	+0.2617993878	+0.1983471505	−0.0634522373
θ_2	+0.2617993878	+0.4299194734	+0.1681200856
θ_3	+0.5235987756	+0.3690547069	−0.1545440687

with every motion cycle being 10 seconds. The simulation results without considering drift-free criterion [i.e., $\lambda = 0$ in (2.2)] are shown in Fig. 8.1 and Table 8.1. The figure illustrates the motion trajectories of the robot arm and the corresponding transients of joint variables. As shown in Fig. 8.1, all joint motion trajectories are not closed, though the end-effector successfully tracks the desired Lissajous path. In other words, the robot final state does not coincide with its initial state [i.e., $\theta_1(10) \neq \theta_1(0)$, $\theta_2(10) \neq \theta_2(0)$, and $\theta_3(10) \neq \theta_3(0)$]. Table 8.1 further shows the

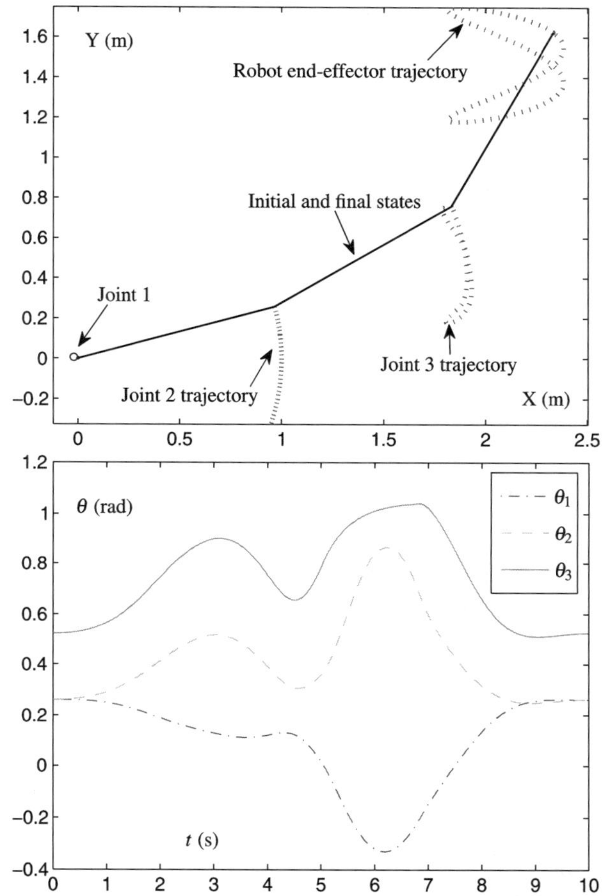

Fig. 8.2 Repetitive motion trajectories of three-link planar robot arm when its end-effector tracking a Lissajous path with the drift-free criterion considered [i.e., $\lambda = 4$ in (2.2)]

joint values and changes. Evidently, the final joint-angle drift is too large to be accepted in practice (especially, in the industrial applications of repetitive motion).

For comparison, we present another computer simulation by considering the drift-free criterion (2.2) with $\lambda = 4$. The simulation results are shown in Figs. 8.2 and 8.3 and Table 8.2. As seen from Fig. 8.2, the resolution scheme could remedy the drift of all robot-joints, and the final-state values fit well with the initial ones after a cycle of the manipulator/end-effector motion. Table 8.2 further shows the corresponding joint-values and final changes in this simulation, wherein the maximal joint-change is less than 0.11057×10^{-4} rad, very tiny and much more acceptable in practice. This substantiates the efficacy of the presented drift-free redundancy-resolution scheme (2.2)–(2.5) on robots tracking Lissajous paths. Moreover, in Figs. 8.1 through 8.3, all joint angles and joint velocities have been kept within their limited ranges, due to the inclusion of bound constraints (2.3) and (2.4) in the RMP scheme (2.2)–(2.5).

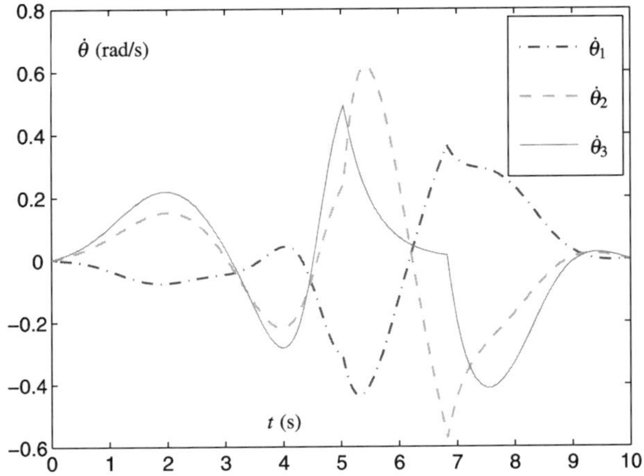

Fig. 8.3 Corresponding joint-velocity profiles when the robot end-effector tracking a Lissajous path with the drift-free criterion considered [i.e., $\lambda = 4$ in (2.2)]

Table 8.2 Joint displacement of three-link robot arm when its end-effector tracking a Lissajous path with the drift-free criterion considered [i.e., $\lambda = 4$ in (2.2)]

Joint	$\theta(0)$ (rad)	$\theta(10)$ (rad)	$\theta(10) - \theta(0)$
θ_1	+0.2617993878	+0.2618026033	$+0.03215 \times 10^{-4}$
θ_2	+0.2617993878	+0.2617961058	-0.03282×10^{-4}
θ_3	+0.5235987756	+0.5235877179	-0.11057×10^{-4}

8.1.2 Triangular Path Tracking

Now, in this part, the end-effector of the three-link planar robot arm is expected to move along a triangular path. The duration of the path-following task at every motion cycle is 27 s (with 9 s for each triangular-side). At every motion cycle, starting from an initial state, the robot arm is expected to return finally to the initial-state configuration.

First, the inverse-kinematics problem with a triangular path following task is handled without considering the drift-free criterion (i.e., $\lambda = 0$). The motion trajectories of the robot and the corresponding profiles of joint variables are depicted in Fig. 8.4. As seen from the figure, all joint-trajectories are evidently not closed (though the end-effector tracks the desired triangular path well). For comparison, Fig. 8.5 shows that the final-state values could be the same as the initial ones when the drift-free criterion (2.2) is considered (e.g., $\lambda = 2$). This substantiates again the efficacy of the presented physically constrained RMP scheme (2.2)–(2.5) on manipulator joint-angle drift remedy.

8.1.3 Elliptical Path Tracking

As the third illustrative simulation example, in this part, we test the three-link planar robot arm moving along an elliptical path. The task duration is 10 s. Similar

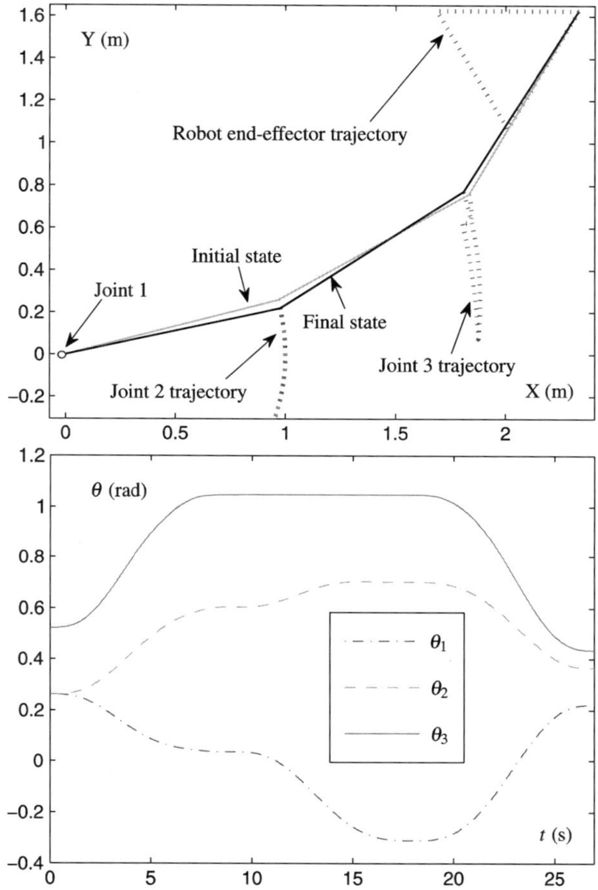

Fig. 8.4 Nonrepetitive motion trajectories of three-link planar robot arm when its end-effector tracking a triangular path without considering the drift-free criterion [i.e., $\lambda = 0$ in (2.2)]

to the aforementioned two examples in this section, (i) without considering the drift-free criterion, all joint-trajectories might not be closed, as depicted in Fig. 8.6; for comparison, and (ii) the final state could match the initial state perfectly if the drift-free criterion (2.2) is considered (e.g., $\lambda = 4$), which is shown affirmably in Fig. 8.7.

In summary, in Sect. 8.1, as illustrated and compared with three types of path-following examples, the joint-angle drift problem of the three-link planar robot arm has been solved in general. The solution results from a physically constrained RMP scheme [i.e., (2.2)–(2.5)], with both drift-free criterion and physical limits considered. A series of computer simulations have demonstrated the efficacy of such a joint-angle drift remedy technique.

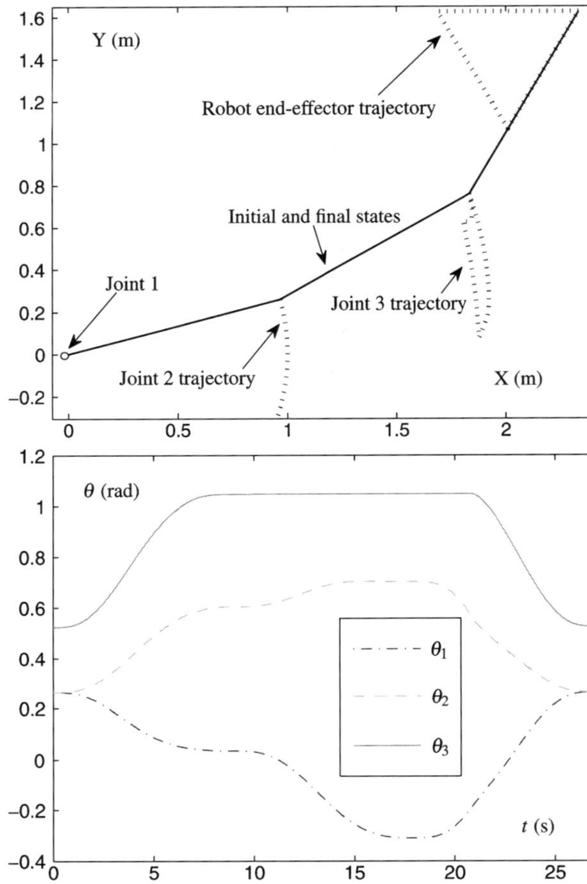

Fig. 8.5 Repetitive motion trajectories of three-link planar robot arm when its end-effector tracking a triangular path with the drift-free criterion considered [i.e., $\lambda = 2$ in (2.2)]

8.2 Four-Link Planar Robot Manipulator

In this section, to further demonstrate the effectiveness of such an RMP scheme, a four-link planar arm is tested to perform square end-effector trajectories through computer simulations [2].

Here, the RMP scheme (2.2)–(2.5) and the LVI-PDNN (5.35) are simulated based on the four-link redundant planar robot manipulator. The end-effector of such a robot arm is expected to track the square path, with task duration of 80 s.

Specifically, the four-link planar robot arm has four DOF in the two-dimensional operational plane, with two degrees of redundancy. Its joint physical limits are given as $\theta^+ = -\theta^- = [\pi/3, \pi/3, \pi/3, \pi/3]^{\mathrm{T}}$ rad and $\dot{\theta}^+ = -\dot{\theta}^- = [\pi, \pi, \pi, \pi]^{\mathrm{T}}$ rad/s. In addition, the initial state $\theta(0) = [\pi/12, \pi/6, -\pi/4, \pi/8]^{\mathrm{T}}$ rad. Note that, in the simulations, all joint angles and joint velocities are kept within their mechanical ranges, due to the inclusion of bound constraint $\xi^- \leqslant \dot{\theta} \leqslant \xi^+$ in the RMP scheme (2.2)–(2.5).

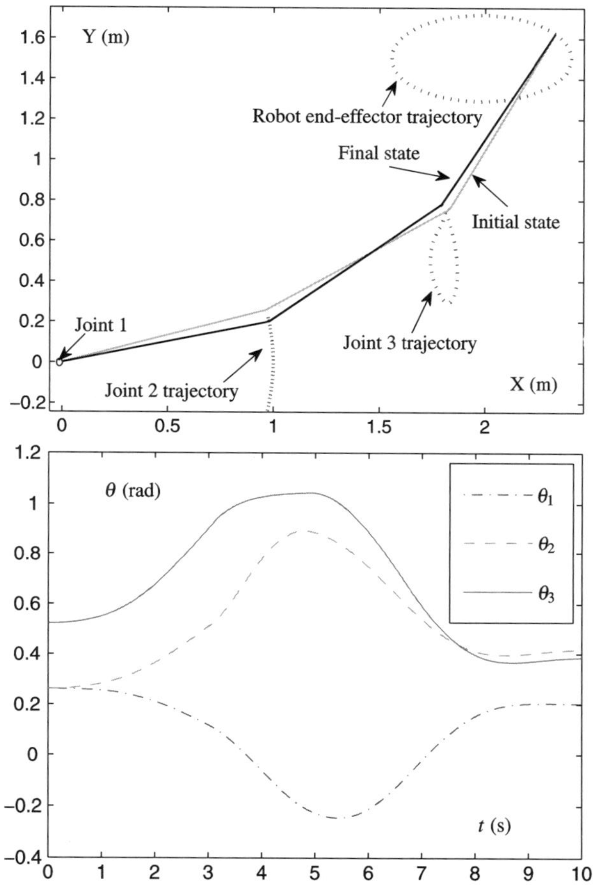

Fig. 8.6 Nonrepetitive motion trajectories of three-link planar robot arm when its end-effector tracking an elliptical path without considering the drift-free criterion [i.e., $\lambda = 0$ in (2.2)]

First, the four-link planar robot arm is expected to track a square path, which is synthesized without considering the drift-free criterion (i.e., $\lambda = 0$). The simulation results are shown in Fig. 8.8 and Table 8.3. Figure 8.8 illustrates the motion trajectories and joint-variable transients of the robot arm when its end-effector is tracking the desired path. As shown in this figure, all joint motion trajectories are not closed, though the end-effector of the manipulator follows the desired square path successfully. In other words, the final state does not coincide with its initial state [i.e., $\theta_1(80) \neq \theta_1(0)$, $\theta_2(80) \neq \theta_2(0)$, $\theta_3(80) \neq \theta_3(0)$, and $\theta_4(80) \neq \theta_4(0)$]. In addition, Table 8.3 shows the corresponding joint-position changes in terms of Cartesian coordinates. Evidently, the final joint-position change $P(80) - P(0)$ of θ_2, θ_3, and θ_4 are too large to be accepted in repetitive-motion tasks. Simply put, this simulation shows us a joint-angle drift problem visibly.

For comparison, as the second computer simulation in this four-link example, the joint-angle drift problem can be solved readily by considering the drift-free criterion (i.e., with $\lambda = 4$). The simulation results are shown in Fig. 8.9 and Table 8.4. As seen from Fig. 8.9, all joints finally return to their initial states. Table 8.4 shows the cor-

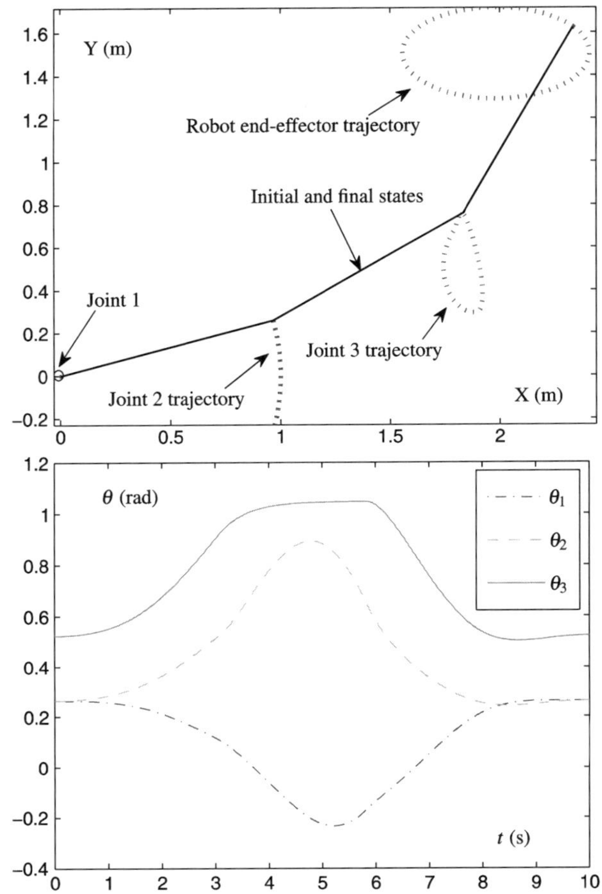

Fig. 8.7 Repetitive motion trajectories of three-link planar robot arm when its end-effector tracking an elliptical path with the drift-free criterion considered [i.e., $\lambda = 4$ in (2.2)]

responding joint-position changes in Cartesian coordinates, wherein the maximal position change is less than 2.51640×10^{-5} m, very tiny and acceptable in practice. This acceptance is in view of the fact that the simulation was performed on a limited-accuracy digital computer with limited memory. As shown additionally in Fig. 8.10, the end-effector position error is very tiny as well, wherein $\tilde{e}_X$ and $\tilde{e}_Y$ denote the components of the position error $\tilde{e}(t) := r(t) - f(\theta(t))$, respectively, along the X- and Y-axes in the base frame of the robot system. This substantiates the efficacy of the authors' RMP scheme (2.2)–(2.5) on repetitive motion planning of robot manipulators.

In summary, in Chap. 2, we has presented a physically constrained RMP scheme (2.2)–(2.5) for redundant robot manipulators, which could incorporate the avoidance of joint physical limits as well. In this section, computer-simulation results based on a four-link planar robot arm tracking a square path have substantiated again the efficacy of such a physically constrained RMP scheme.

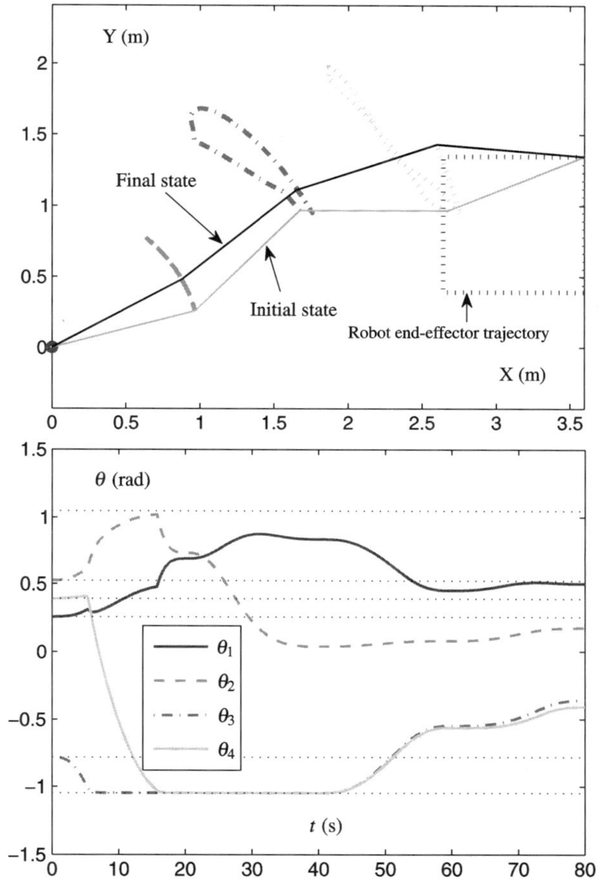

Fig. 8.8 Nonrepetitive motion trajectories of four-link planar robot arm when its end-effector tracking a square path synthesized without considering the drift-free criterion [i.e., $\lambda = 0$ in (2.2)]

Table 8.3 Joint-position change (m) of four-link robot arm when its end-effector tracking a square path without considering the drift-free criterion [i.e., $\lambda = 0$ in (2.2)]

#	$P(0)$	$P(80)$	$P(80) - P(0)$
Joint 1	(+0.0000, +0.0000)	(+0.0000, +0.0000)	(+0.0000, +0.0000)
Joint 2	(+0.9659, +0.2588)	(+0.8762, +0.4819)	(−0.0897, +0.2230)
Joint 3	(+1.6730, +0.9659)	(+1.6523, +1.1124)	(−0.0206, +0.1465)
Joint 4	(+2.6730, +0.9659)	(+2.6003, +1.4308)	(−0.0727, +0.4648)
End-effector	(+3.5969, +1.3486)	(+3.5969, +1.3486)	(+0.0000, +0.0000)

8.3 Five-Link Planar Robot Manipulator

The five-link planar robot manipulator has five degrees of freedom (or to say, five joints). In this section, both LVI-PDNN (5.35) and DNN (4.26)–(4.27) are applied to the online joint-angle drift-free redundancy resolution of the five-link planar robot

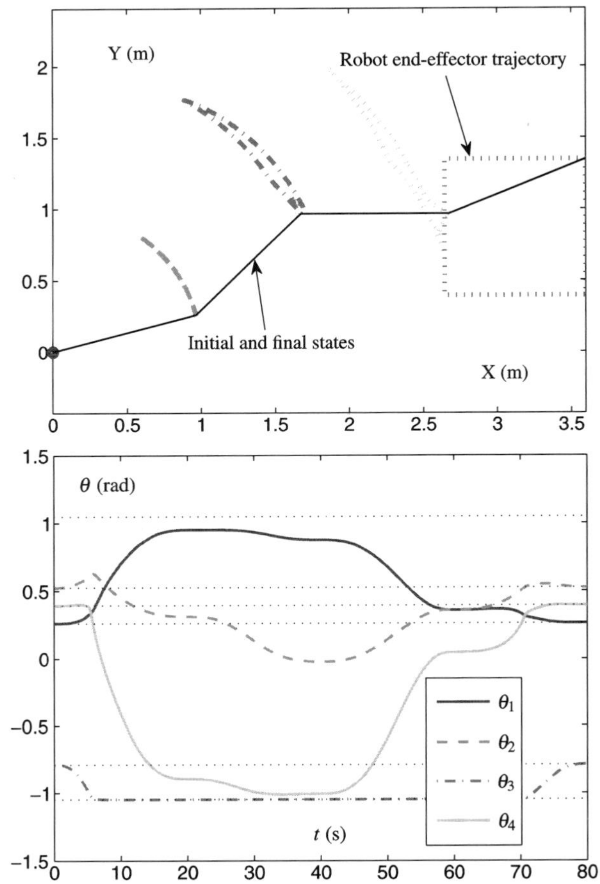

Fig. 8.9 Repetitive motion trajectories of four-link planar robot arm when its end-effector tracking a square path with the drift-free criterion considered [i.e., $\lambda = 4$ in (2.2)]

Table 8.4 Joint-position change (m) of four-link robot arm when its end-effector tracking a square path with the drift-free criterion considered [i.e., $\lambda = 4$ in (2.2)]

#	$P(0)$	$P(80)$	$P(80) - P(0)$
Joint 1	(+0.0000, +0.0000)	(+0.0000, +0.0000)	(+0.0000, +0.0000)
Joint 2	(+0.9659, +0.2588)	(+0.9659, +0.2588)	$(+0.1832 \times 10^{-5}, -0.6837 \times 10^{-5})$
Joint 3	(+1.6730, +0.9659)	(+1.6731, +0.9659)	$(+2.0159 \times 10^{-5}, -2.5164 \times 10^{-5})$
Joint 4	(+2.6730, +0.9659)	(+2.6731, +0.9659)	$(+2.0159 \times 10^{-5}, -1.0386 \times 10^{-5})$
End-effector	(+3.5969, +1.3486)	(+3.5969, +1.3486)	$(+1.7053 \times 10^{-5}, -0.2889 \times 10^{-5})$

manipulator, which substantiates also the effectiveness of the physically constrained RMP scheme [3].

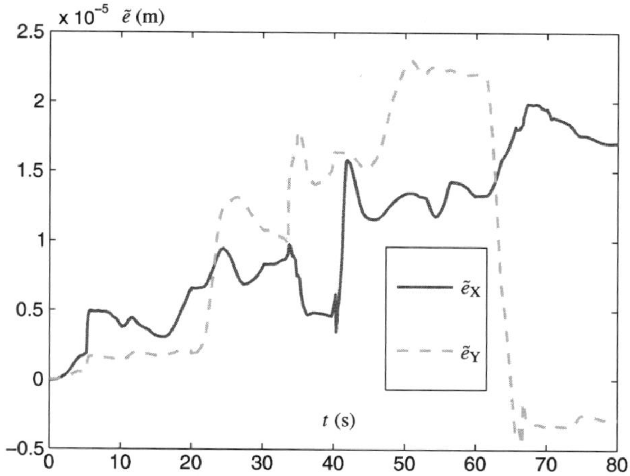

Fig. 8.10 End-effector position error of four-link planar robot arm when tracking a square path synthesized with the drift-free criterion considered [i.e., $\lambda = 4$ in (2.2)]

8.3.1 Simulations with LVI-PDNN Solver

In this subsection, the RMP scheme (2.2)–(2.5) is simulated and applied based on the five-link redundant planar robot arm [2]. The end-effector of such a robot arm is expected to track a square, with task duration being 80 s. The five-link planar robot arm operates in the two-dimensional plane with three-DOF redundancy. The physical limits of joint angles and joint velocities are given as $\theta^+ = -\theta^- = [\pi/3, \pi/3, \pi/3, \pi/3, \pi/3]^{\mathrm{T}}$ rad and $\dot{\theta}^+ = -\dot{\theta}^- = [\pi, \pi, \pi, \pi, \pi]^{\mathrm{T}}$ rad/s, respectively. In addition, the initial state $\theta(0) = [-\pi/12, \pi/12, \pi/6, -\pi/4, \pi/3]^{\mathrm{T}}$ rad. As seen from Fig. 8.11 and Table 8.5, closed trajectories may not be obtained in the joint space when we do not consider the drift-free criterion (i.e., with $\lambda = 0$), though the end-effector tracks the desired square path in the work-plane successfully. In contrast, Fig. 8.12 and Table 8.6 show that the final state could fit well with the initial state by using the drift-free criterion (i.e., with $\lambda = 4$), wherein the final joint-position change is very tiny, less than 10^{-5} m (i.e., 10^{-2} in millimeters). Simulation results based on this five-link planar robot arm also substantiate well the efficacy of RMP scheme (2.2)–(2.5).

In summary, this part has tested the RMP scheme (2.2)–(2.5) for a redundant robot manipulator, which has incorporated the avoidance of joint physical limits as well. Computer-simulation results based on the five-link planar robot arm tracking a square path have substantiated again the efficacy of such an RMP scheme.

8.3.2 Simulations with DNN Solver

In this subsection, we apply the dual neural network [with design parameter $\bar{\gamma} = 10^8$ in (4.26) in Chap. 4] to the online drift-free redundancy resolution of the same five-link manipulator [4]. Computer simulation is performed for the path-following task that the end-effector tracks a circle with radius 0.4426 m and task duration 10 s. In

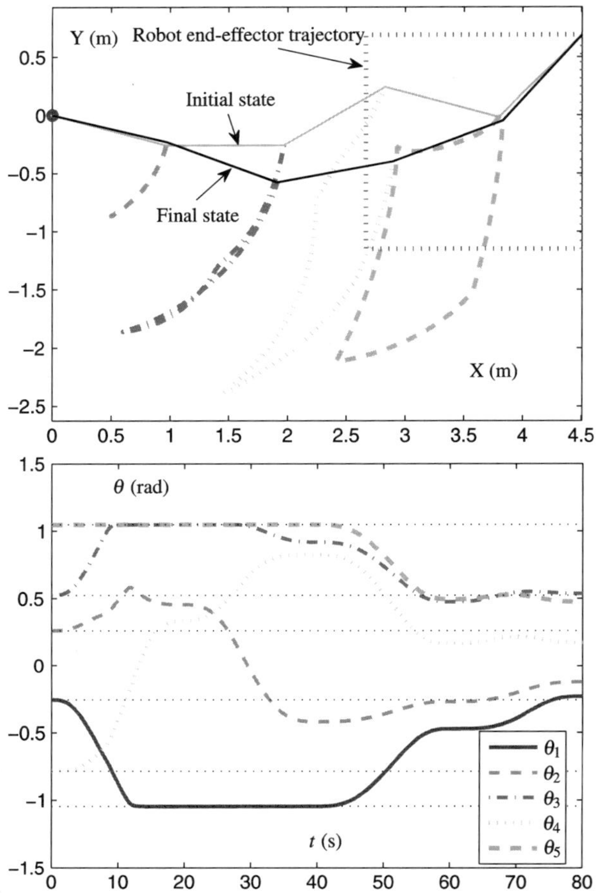

Fig. 8.11 Nonrepetitive motion trajectories of five-link planar robot arm when its end-effector tracking a square path synthesized without considering the drift-free criterion [i.e., $\lambda = 0$ in (2.2)]

Table 8.5 Joint-position change (m) of five-link robot arm when its end-effector tracking a square path without considering the drift-free criterion [i.e., $\lambda = 0$ in (2.2)]

#	$P(0)$	$P(80)$	$P(80) - P(0)$
Joint 1	(+0.0000, +0.0000)	(+0.00000, +0.0000)	(+0.0000, +0.0000)
Joint 2	(+0.9659, −0.2588)	(+0.9728, −0.2315)	(+0.0069, +0.0273)
Joint 3	(+1.9659, −0.2588)	(+1.9104, −0.5794)	(−0.0555, −0.3206)
Joint 4	(+2.8320, +0.2412)	(+2.8939, −0.3985)	(+0.0619, −0.6397)
Joint 5	(+3.7979, −0.0176)	(+3.8309, −0.0491)	(+0.0330, −0.0315)
End-effector	(+4.5050, +0.6895)	(+4.5050, +0.6895)	(+0.0000, −0.0000)

this study, only the position of the end-effector is considered, the Jacobian matrix is thus 2×5 in dimension, and the degrees of redundancy are $5 - 2 = 3$. The initial joint state $\theta(0) = [\pi/12, \pi/12, \pi/6, \pi/3, \pi/2]^{\mathrm{T}}$ rad, which is expected to be the final state of the robot. Here, we limit the joint vector (and joint-velocity vector) be-

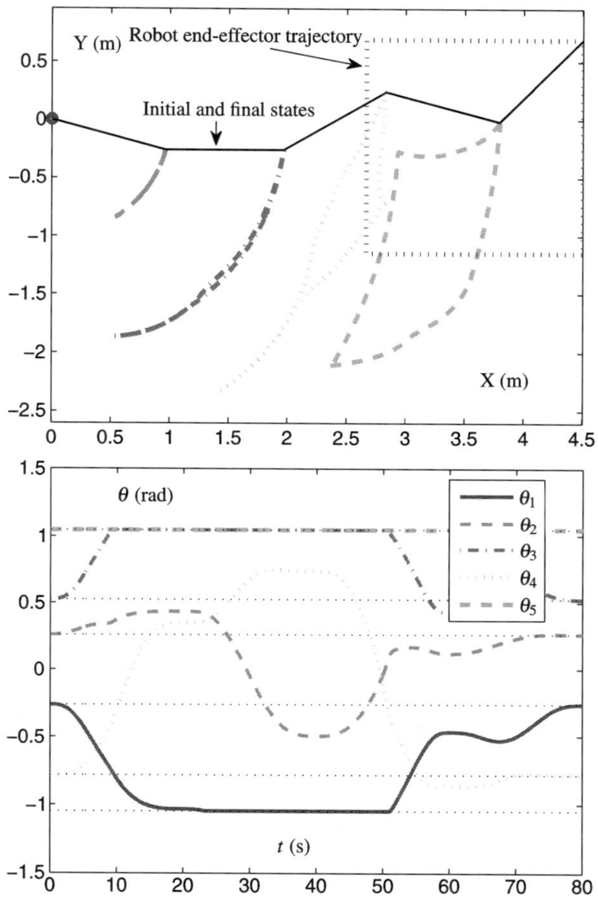

Fig. 8.12 Repetitive motion trajectories of five-link planar robot arm when its end-effector tracking a square path with the drift-free criterion considered [i.e., $\lambda = 4$ in (2.2)]

Table 8.6 Joint-position change (m) of five-link robot arm when its end-effector tracking a square path with the drift-free criterion considered [i.e., $\lambda = 4$ in (2.2)]

#	$P(0)$	$P(80)$	$P(80) - P(0)$
Joint 1	(+0.0000, +0.0000)	(+0.0000, +0.0000)	(+0.0000, +0.0000)
Joint 2	(+0.9659, −0.2588)	(+0.9659, −0.2588)	$(-0.1528 \times 10^{-5}, -0.5704 \times 10^{-5})$
Joint 3	(+1.9659, −0.2588)	(+1.9659, −0.2588)	$(-0.1528 \times 10^{-5}, -0.8643 \times 10^{-5})$
Joint 4	(+2.8320, +0.2412)	(+2.8319, +0.2412)	$(-0.3636 \times 10^{-5}, -0.4993 \times 10^{-5})$
Joint 5	(+3.7979, −0.0176)	(+3.7979, −0.0176)	$(-0.2177 \times 10^{-5}, +0.0453 \times 10^{-5})$
End-effector	(+4.5050, +0.6895)	(+4.5050, +0.6895)	$(-0.0782 \times 10^{-5}, -0.0942 \times 10^{-5})$

tween $\pm[3\pi/4, 3\pi/4, 3\pi/4, 3\pi/4, 3\pi/4]^{\mathrm{T}}$ rad (and $\pm[\pi/3, \pi/3, \pi/3, \pi/3, \pi/3]^{\mathrm{T}}$ rad/s, respectively).

First, we illustrate the redundancy-resolution results synthesized by using the dual neural network (4.26)–(4.27) but without considering drift-free criterion [i.e.,

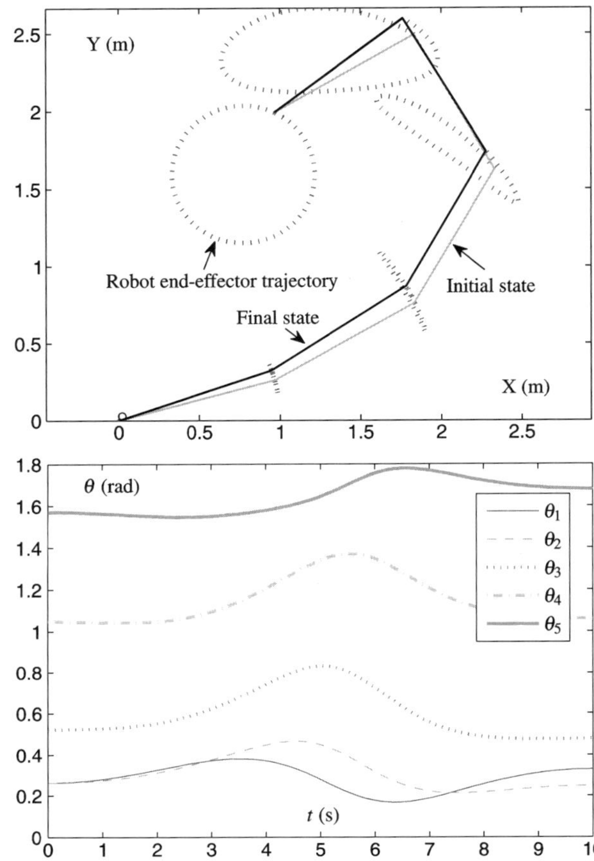

Fig. 8.13 End-effector of five-link planar manipulator moving along a circle without considering the drift-free criterion [i.e., $\lambda = 0$ in (2.2)]

Table 8.7 Joint displacement of five-link planar robot arm when its end-effector tracking a circle without considering the drift-free criterion [i.e., $\lambda = 0$ in (2.2)]

Joint #	$\theta(0)$ (rad)	$\theta(10)$ (rad)	$\theta(10) - \theta(0)$
θ_1	+0.2617993877	+0.3297592033	+0.0679598155
θ_2	+0.2617993877	+0.2483077534	−0.0134916343
θ_3	+0.5235987755	+0.4762764779	−0.0473222976
θ_4	+1.0471975511	+1.0583832174	+0.0111856662
θ_5	+1.5707963267	+1.6794311576	+0.1086348308

$\lambda = 0$ and thus $\hat{c} = 0$ in (2.2)]. Figure 8.13 illustrates the motion trajectories of the five-link planar manipulator when its end-effector moves along a circle, which are accompanied by the corresponding transients of joint variables as shown in its lower graph. As seen from Fig. 8.13, the solution is not repetitive in the sense that the final state of the five-link planar manipulator does not coincide with its initial state. The joint displacements are shown in Table 8.7, of which the maximal one

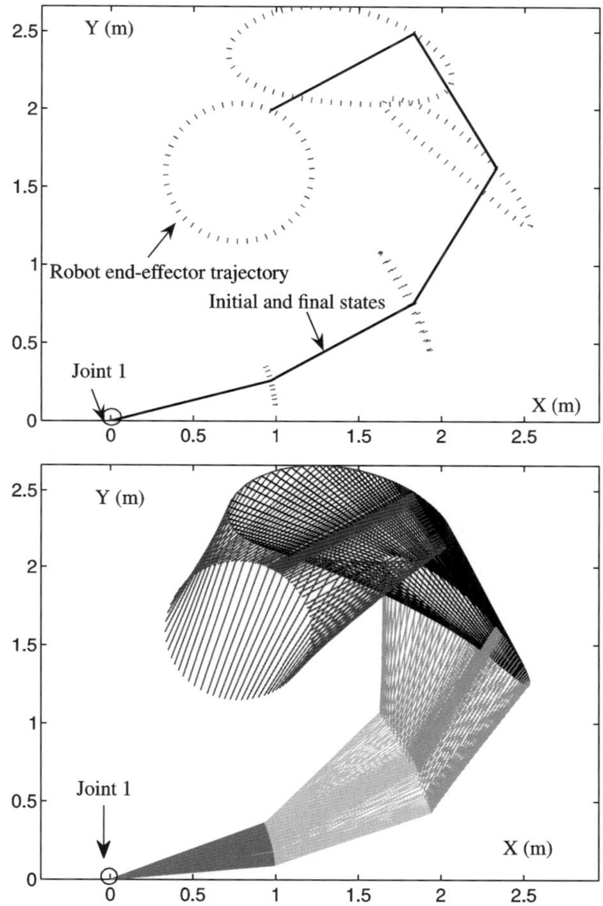

Fig. 8.14 End-effector of five-link planar manipulator moving repetitively along a circle with the drift-free criterion considered [i.e., $\lambda = 6$ in (2.2)]

is 0.1 rad (i.e., 5.7°). Hence, an inefficient and undesirable readjustment might be needed in the repetitive-motion control context if we accept so. Simply speaking, we encountered a joint-angle drift problem here.

Secondly, for comparison with the preceding simulation and for illustration of the effectiveness of the physically constrained RMP scheme, the above inverse-kinematics problem is solved again but with $\lambda = 6$ and $\mu = 4$ (i.e., in the physically constrained RMP manner). As seen from Fig. 8.14, the solution becomes repetitive in the sense that the final and initial states of such a five-link planar robot manipulator coincide with each other very well. Figure 8.15 shows the profiles of joint-angle and joint-velocity variables, which, evidently, all have been kept within their physical ranges. Figure 8.16 tells that the maximal position error is less than 1.5×10^{-5} m and that the maximal end-effector velocity error is less than 1.5×10^{-5} m/s. Here, $\tilde{e}_X$ and $\tilde{e}_Y$ denote again the components of the position error $\tilde{e}(t) := r(t) - f(\theta(t))$ respectively along the X- and Y-axes in the base frame of the robot system. Similarly, $\dot{\tilde{e}}_X$ and $\dot{\tilde{e}}_Y$ denote the corresponding components

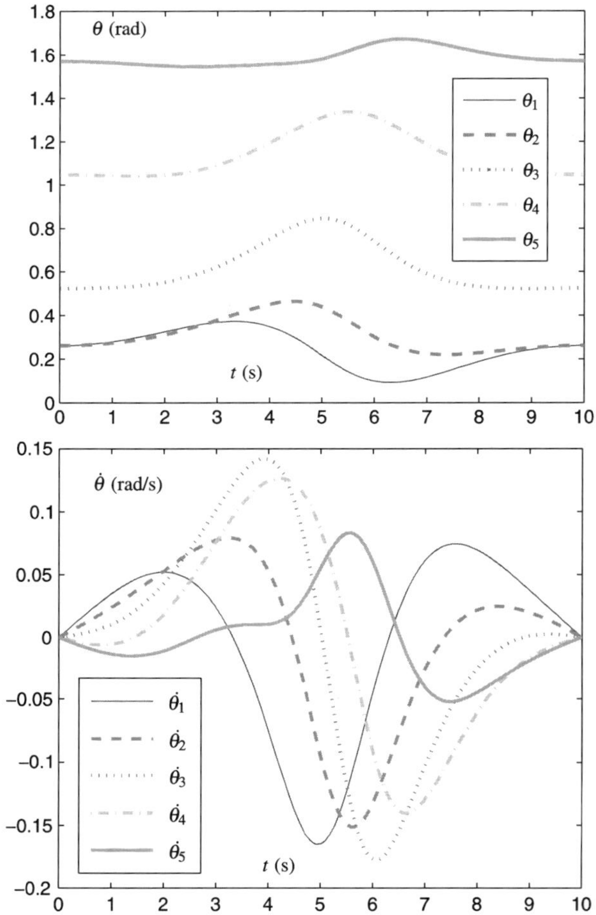

Fig. 8.15 Joint profiles of five-link planar robot manipulator when its end-effector moving along a circle with the drift-free criterion considered [i.e., $\lambda = 6$ in (2.2)]

of velocity error $\dot{e}(t) := \dot{r}(t) - J(\theta(t))\dot{\theta}(t)$. Besides, Table 8.8 lists the joint displacements generated in this simulation. Such displacement results are acceptably perfect in practice (with displacements being around 10^{-6} rad and the maximal one being less than 4.5×10^{-6} rad $= 2.58 \times 10^{-4}$ deg). This tiny displacement is understandable in view of the fact that the simulation has to be performed generally on limited-accuracy limited-memory digital computers.

The above computer-simulation results based on the five-link planar robot have substantiated the efficacy of the presented dual neural network approach to online drift-free redundancy resolution of physically constrained robot manipulators.

8.4 Six-Link Planar Robot Manipulator

In this section, the most preferred quadratic-form scheme and the LVI-PDNN are employed for RMP of a six-link planar arm with joint physical limits considered

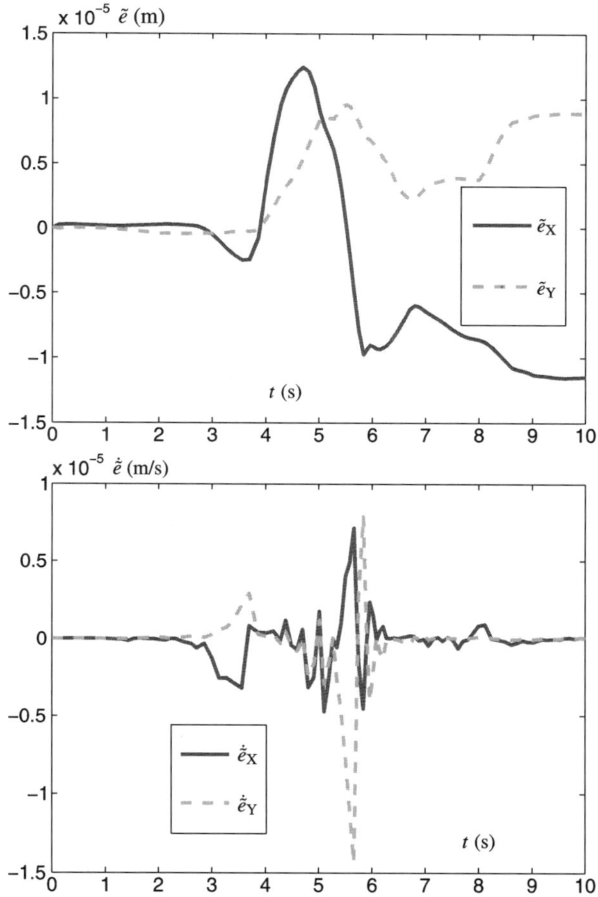

Fig. 8.16 End-effector position error and velocity error of five-link planar robot manipulator moving along a circle with the drift-free criterion considered [i.e., $\lambda = 6$ in (2.2)]

Table 8.8 Joint displacement of five-link planar robot arm when its end-effector tracking a circle with the drift-free criterion considered [i.e., $\lambda = 6$ in (2.2)]

Joint #	$\theta(0)$ (rad)	$\theta(10)$ (rad)	$\theta(10) - \theta(0)$
θ_1	+0.2617993877	+0.2617953407	-4.04702×10^{-6}
θ_2	+0.2617993877	+0.2617975279	-1.85986×10^{-6}
θ_3	+0.5235987755	+0.5235988849	$+0.10931 \times 10^{-6}$
θ_4	+1.0471975511	+1.0471996754	$+2.12422 \times 10^{-6}$
θ_5	+1.5707963267	+1.5707985805	$+2.25377 \times 10^{-6}$

[2]. The end-effector's square path is tested through computer simulations. The task duration is 80 s.

Specifically, the six-link planar robot arm has six DOF, with four degrees of redundancy. It is employed as well to verify the effectiveness and efficiency of

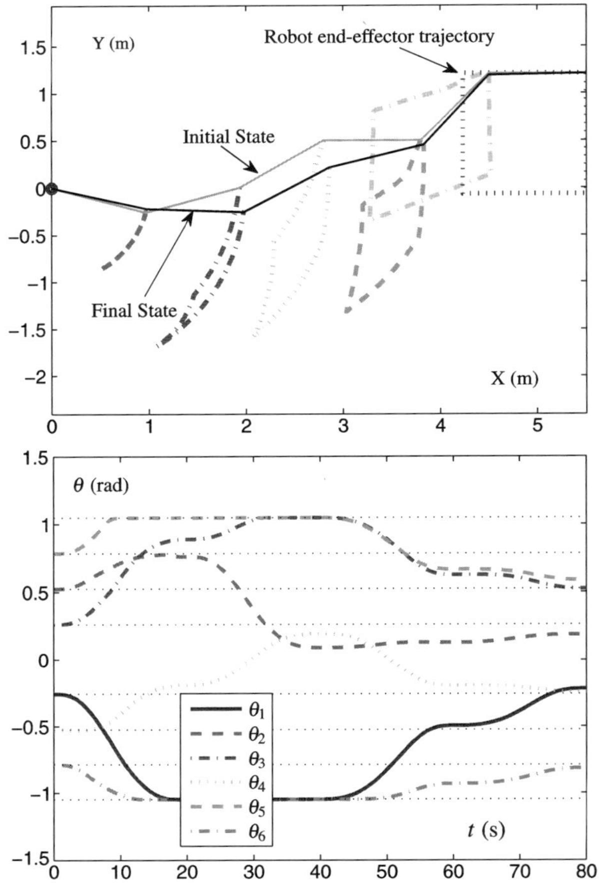

Fig. 8.17 Nonrepetitive motion trajectories of six-link planar robot arm when its end-effector tracking a square path synthesized without considering the drift-free criterion [i.e., $\lambda = 0$ in (2.2)]

the presented physically constrained RMP scheme (2.2)–(2.5). Similar to previous examples, the limits of all joint angles and velocities are given as $\pm\pi/3$ rad and $\pm\pi$ rad/s, respectively. It is worth mentioning that the initial state $\theta(0) = [-\pi/12, \pi/6, \pi/12, -\pi/6, \pi/4, -\pi/4]^{\mathrm{T}}$ rad. Figures 8.17 and 8.18 illustrate the nonrepetitive and repetitive motion trajectories of such a six-link planar robot arm synthesized via the RMP scheme (2.2)–(2.5). On one hand, Fig. 8.17 shows that, without considering the drift-free criterion, some joint-trajectories are not closed in view of the fact that $\theta_2(80) \neq \theta_2(0)$, $\theta_3(80) \neq \theta_3(0)$, $\theta_4(80) \neq \theta_4(0)$, and $\theta_5(80) \neq \theta_5(0)$. On the other hand, as seen from Fig. 8.18, all final joint states coincide well with their initial ones by using RMP scheme (2.2)–(2.5) (specifically, using the drift-free criterion with $\lambda = 4$).

In summary, this section has tested the physically constrained RMP scheme (2.2)–(2.5) for a six-link robot manipulator, which could incorporate the avoidance of joint physical limits as well. Computer-simulation results tracking a square path have substantiated again the efficacy of such an RMP scheme.

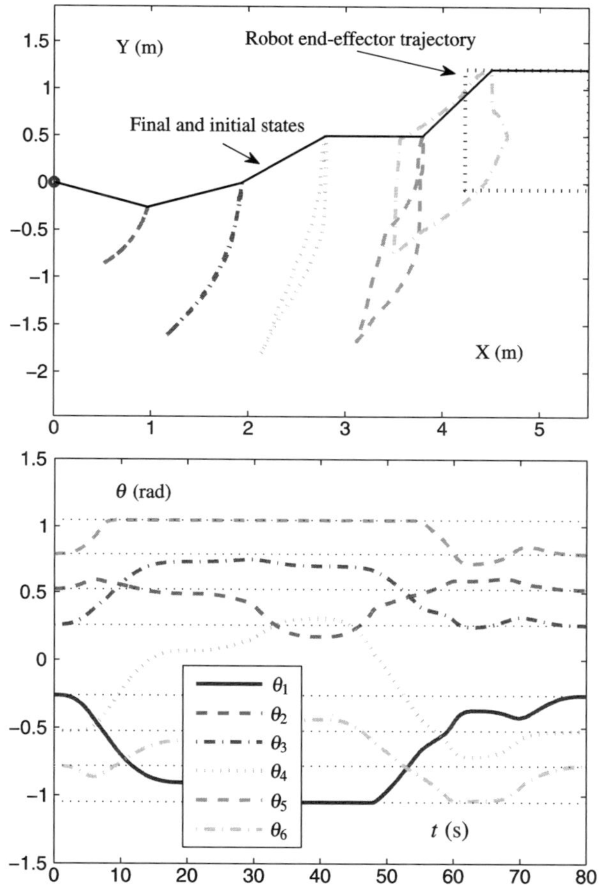

Fig. 8.18 Repetitive motion trajectories of six-link planar robot arm when its end-effector tracking a square path with the drift-free criterion considered [i.e., $\lambda = 4$ in (2.2)]

8.5 Chapter Conclusions

This chapter has applied the physically constrained RMP scheme (2.2)–(2.5) to three planar robot manipulators, i.e., a three-link planar robot manipulator, a four-link planar robot manipulator, and a six-link planar robot manipulator. For illustrating the efficacy of the presented RMP scheme, LVI-PDNN and DNN are both employed to solve the resultant QP problem. Computer-simulation results have substantiated the efficacy of such a physically constrained RMP scheme.

References

1. Zhang Y, Li X, Zhu H, Tan N (2009) Joint-angle-drift remedy of three-link planar robot arm performing different types of end-effector trajectories. In: Proceedings of IEEE international conference on intelligent computing and intelligent systems, pp 581–585

2. Chen K, Zhang L, Zhang Y (2008) Cyclic motion generation of multi-link planar robot performing square end-effector trajectory analyzed via gradient-descent and Zhang et al's neural-dynamic methods. In: Proceedings of the 2nd international symposium on systems and control in aerospace and astronautics, pp 1–6
3. Zhang Y, Tan Z, Chen K, Yang Z, Lv X (2009) Repetitive motion of redundant robots planned by three kinds of recurrent neural networks and illustrated with a four-link planar manipulator's straight-line example. Robot Auton Syst 57(6–7):645–651
4. Zhang Y, Tan Z, Yang Z, Lv X (2008) A dual neural network applied to joint angle drift-free resolution of five-link planar robot arm. In: Proceedings of IEEE international conference on information and automation, pp 1274–1279

Chapter 9
PUMA560 Examples

Abstract In this chapter, the physically constrained RMP scheme (2.2)–(2.5) is applied to the PUMA560 robot manipulator by using two QP solvers, i.e., the LVI-PDNN (5.35) and the simplified LVI-PDNN (5.36). Different desired motion paths of the PUMA560 end-effector (i.e., circular and straight-line paths) are tested for illustrative and comparative purposes.

9.1 Simulations Using LVI-PDNN Solver

In this section, physically constrained RMP scheme (2.2)–(2.5) is simulated based on the PUMA560 robot manipulator and LVI-PDNN (5.35) [1]. In the simulations, the following limited joint ranges of the PUMA560 robot manipulator are used: $[-2.984, 2.984]$, $[-3.378, 0.807]$, $[-0.974, 3.378]$, $[-2.064, 3.190]$, $[-1.877, 0.038]$, $[-3.378, 3.378]$ rad respectively for θ_1 through θ_6. In addition, the limited range for joint velocity $\dot{\theta}$ is $[-1.5, 1.5]^6$ rad/s. The PUMA560 end-effector is expected to track the circular and straight-line paths via the QP-based RMP scheme (2.2)–(2.5) and LVI-PDNN (5.35). The corresponding simulation results are shown as follows.

9.1.1 Circular Path Tracking

First, we show the inverse-kinematics results of the PUMA560 robot manipulator without considering drift-free criterion. That is, with the drift-free coefficient $\lambda = 0$, the LVI-PDNN is applied to the PUMA560 robot manipulator. The design parameter $\bar{\gamma} = 10^8$. The desired motion of the PUMA560 end-effector is a circular path with radius 0.1 m and the revolute angle about X-axis being $\pi/6$ rad. The motion task duration is 10 s, and the initial joint vector $\theta(0) = [0, 0, 0, 0, 0, 0]^{\mathrm{T}}$ rad. Figure 9.1 illustrates the motion trajectories and joint-variable transients of the PUMA560 manipulator when its end-effector moves along a circle in the three-dimensional workspace. It shows that there is no joint variable exceeding its limited range, because all the physical constraints have been considered. However, the solution is not repetitive in the sense that the final state of the PUMA560 manipulator does

Y. Zhang, Z. Zhang, *Repetitive Motion Planning and Control of Redundant Robot Manipulators*, DOI 10.1007/978-3-642-37518-7_9,

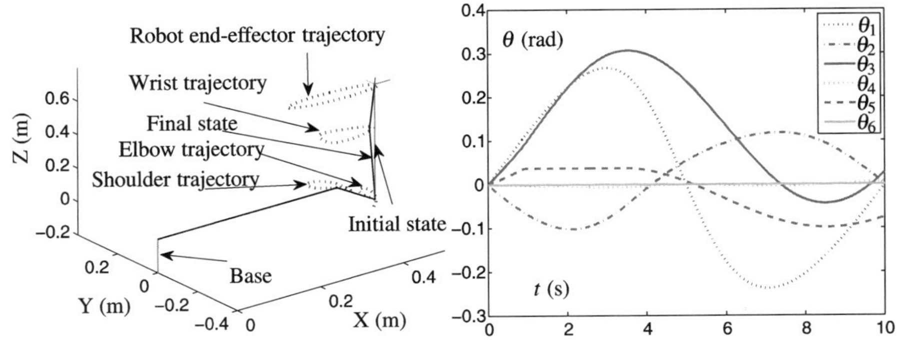

Fig. 9.1 PUMA560 does not return to its initial state after the end-effector completing a circular–path task, which is the so-called joint-angle drift (or to say, nonrepetitive) problem

not coincide with its initial state. Mathematically, $\theta_3(10) \neq \theta_3(0)$, $\theta_4(10) \neq \theta_4(0)$, and $\theta_5(10) \neq \theta_5(0)$. Hence, an additional self-motion readjustment may be needed if repetitive motion control is pursued. This would certainly be less efficient and unwanted in repetitive path-tracking tasks.

Second, for comparison, by another computer simulation, the drift-free criterion with $\lambda = 4$ is exploited. For the same circular-path tracking task, the LVI-PDNN is then applied again to the PUMA560 robot manipulator. As seen from Fig. 9.2, all joint variables have been kept within their limited ranges, and the solution is repetitive in the sense that the initial state and final state of the PUMA560 manipulator are equal. Figure 9.2 also shows that the maximal position error is less than 4×10^{-5} m. Here, $\tilde{e}_X$, $\tilde{e}_Y$, and $\tilde{e}_Z$ denote the components of the position error $\tilde{e}(t) := r(t) - f(\theta(t))$ respectively along the X-, Y-, and Z-axes in the base frame of the robot system.

The circular-path tracking simulations demonstrate the capability of the LVI-PDNN (5.35) for resolving online the drift-free redundancy problem of the physically constrained PUMA560 robot manipulator.

9.1.2 Straight-Line Path Tracking

In this computer-simulation example, the PUMA560 manipulator is expected to move forwards and then backwards along a straight-line segment. The straight-line segment with length being 2.25 m, at every motion cycle, starts with the PAUMA560 initial state $\theta(0) = [0, 0, 0, 0, 0, 0]^{\mathrm{T}}$ rad and will finally return to the initial state. Angles of the desired straight-line segment making with XY, YZ, and ZX planes are respectively 0 rad, $\pi/4$ rad, and $\pi/4$ rad. The duration of the path-tracking task at every motion cycle is 20 s.

The drift-free criterion and LVI-PDNN are then applied to the control of the physically constrained PUMA560 manipulator. In this straight-line example, the design parameters of the LVI-PDNN are selected to be the same as those in the

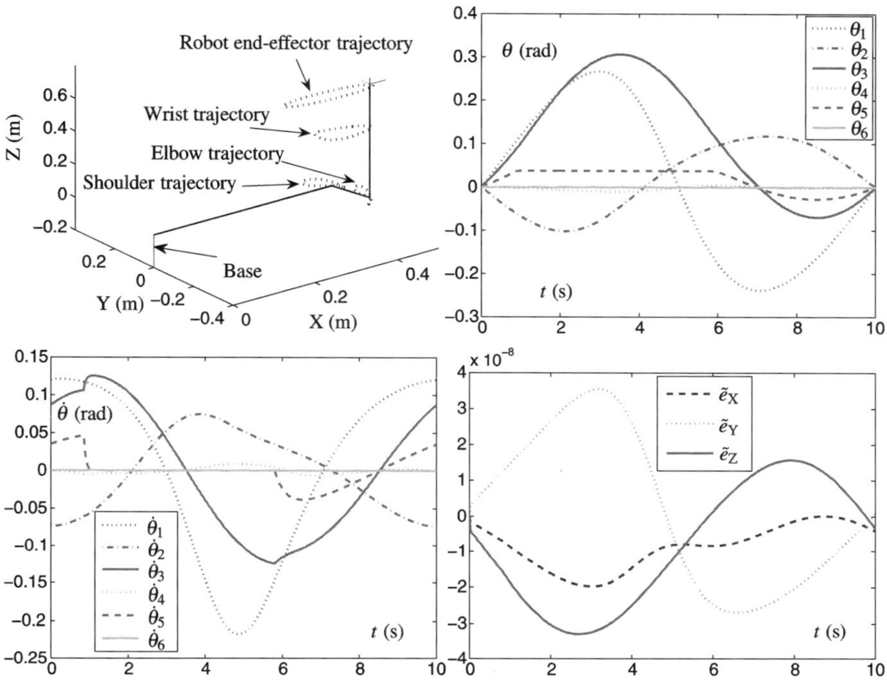

Fig. 9.2 PUMA560 motion trajectories and transients when its end-effector moves along a circle, which is synthesized by the LVI-PDNN

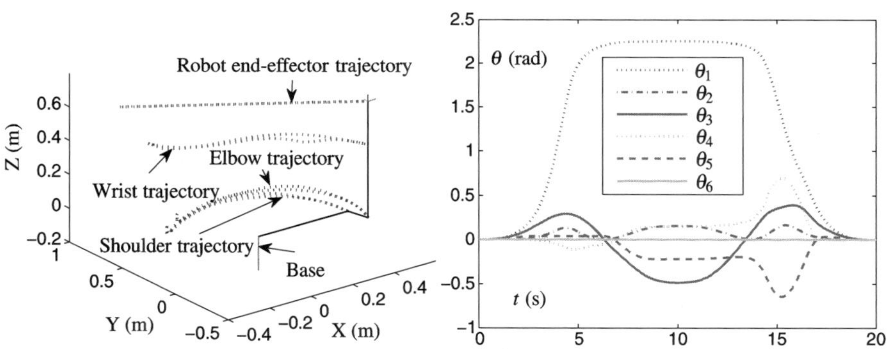

Fig. 9.3 PUMA560 motion trajectories and transients when its end-effector moves forwards and then backwards along a straight-line segment path, which is synthesized by the drift-free criterion and LVI-PDNN

circular path tracking example. As shown in Fig. 9.3, the final state of the PUMA560 manipulator coincides with its initial state. In addition, all joint variables remain within their limited ranges.

Similar to the circular-path tracking example, this straight-line path tracking example also demonstrates the efficacy of the drift-free problem formulation and its online LVI-PDNN solution on the real-time redundancy resolution of physically constrained robot manipulators.

In summary, as illustrated via PUMA560 examples, we have solved the joint-angle drift problem of physically constrained redundant manipulators. The solution is based on a quadratic-programming problem formulation and an LVI-based primal–dual neural network (LVI-PDNN) as the real-time solver. Simulation results based on the PUMA560 robot manipulator have demonstrated the efficacy of the drift-free problem formulation and its neural-network solver on the real-time inverse-kinematics control of joint-constrained redundant manipulators.

9.2 Simulations Using S-LVI-PDNN Solver

In this section, the mentioned physically constrained RMP scheme (2.2)–(2.5) is simulated based on the PUMA560 robot manipulator and the simplified LVI-PDNN (5.36) (i.e., another QP solver abbreviated as S-LVI-PDNN differing from the LVI-PDNN) [2]. Note that the limited joint and joint-velocity ranges used in the simulations for the PUMA560 robot manipulator are set the same as before. The PUMA560 end-effector is expected to track the circular and straight-line paths via the aforementioned RMP scheme and the simplified LVI-PDNN. The corresponding simulation results are shown for illustrative and comparative purposes.

9.2.1 Circular Path Tracking

First, we show the inverse-kinematics results of the PUMA560 robot manipulator without considering drift-free criterion. That is, with drift-free coefficient $\lambda = 0$, the simplified LVI-PDNN (5.36) is applied to the PUMA560 robot manipulator. The design parameter $\bar{\gamma} = 10^8$. The desired motion of the PUMA560 end-effector is a circular path with radius 0.1 m and the revolute angle about X-axis being $\pi/6$ rad. The task duration is 10 s, and the initial joint variables $\theta(0) = [0, 0, 0, 0, 0, 0]^{\mathrm{T}}$ rad. Figure 9.4 illustrates the motion trajectories and joint-variable transients of the PUMA560 robot manipulator when its end-effector moves along a circle in the three-dimensional workspace. It shows that there is no joint variable exceeding its limited range, as all joint physical limits are considered in the scheme formulation. However, the solution is not repetitive in the sense that the final state of the PUMA560 robot manipulator does not coincide with its initial state, of which the final joint-angle values are given in Table 9.1.

Second, for comparison, by another computer simulation, the drift-free criterion with $\lambda = 8$ is exploited. For the same circular-path tracking task, the simplified LVI-PDNN is applied again to controlling the PUMA560 robot manipulator. As seen from Fig. 9.5, all joint variables are kept within their limited ranges, and the solution is repetitive as well in the sense that the initial and final states of the PUMA560

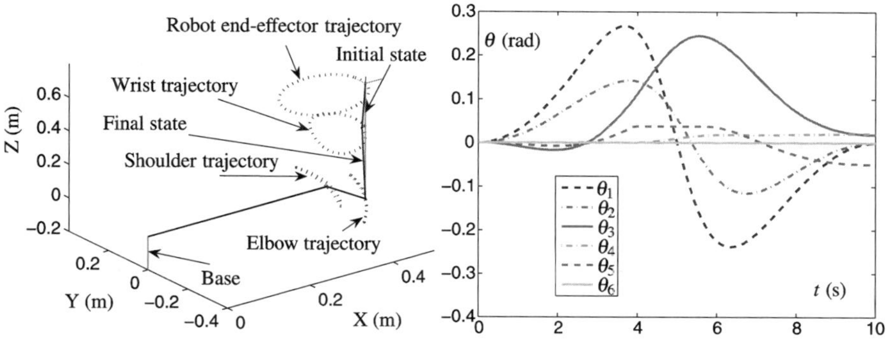

Fig. 9.4 PUMA560 does not return to the initial state after its end-effector completing a circular path tracking task

Table 9.1 Final values of joint variables of PUMA560 robot manipulator after tracking a circle but without considering drift-free criterion

#	joint	$\theta(0)$ (rad)	$\theta(10)$ (rad)	$\theta(10)$ (deg)
1	θ_1	0	−0.00058885721776	−0.03373903331346
2	θ_2	0	−0.00040393758281	−0.02314391868173
3	θ_3	0	+0.01862240756550	+1.06698535787564
4	θ_4	0	+0.02108972921745	+1.20835247523363
5	θ_5	0	−0.04928997254042	−2.82410739888179
6	θ_6	0	0.00000000000000	0.00000000000000

Table 9.2 Final values of joint variables of PUMA560 robot manipulator after tracking a circle with drift-free criterion considered

#	joint	$\theta(0)$ (rad)	$\theta(10)$ (rad)
1	θ_1	0	$-0.00075824894098 \times 10^{-6}$
2	θ_2	0	$+0.00323661561217 \times 10^{-6}$
3	θ_3	0	$-0.04421887315539 \times 10^{-6}$
4	θ_4	0	$+0.19473845684456 \times 10^{-6}$
5	θ_5	0	$+0.11215303890168 \times 10^{-6}$
6	θ_6	0	$0.00000000000000 \times 10^{-6}$

manipulator are very close. Table 9.2 gives the final values of joint variables in this simulation. Note that such a simulation (and solution) is performed on a limited-accuracy digital computer with limited memory.

The circular-path tracking example demonstrates the capability of the simplified LVI-PDNN for online resolution of the drift-free redundancy problem of physically constrained robot manipulator.

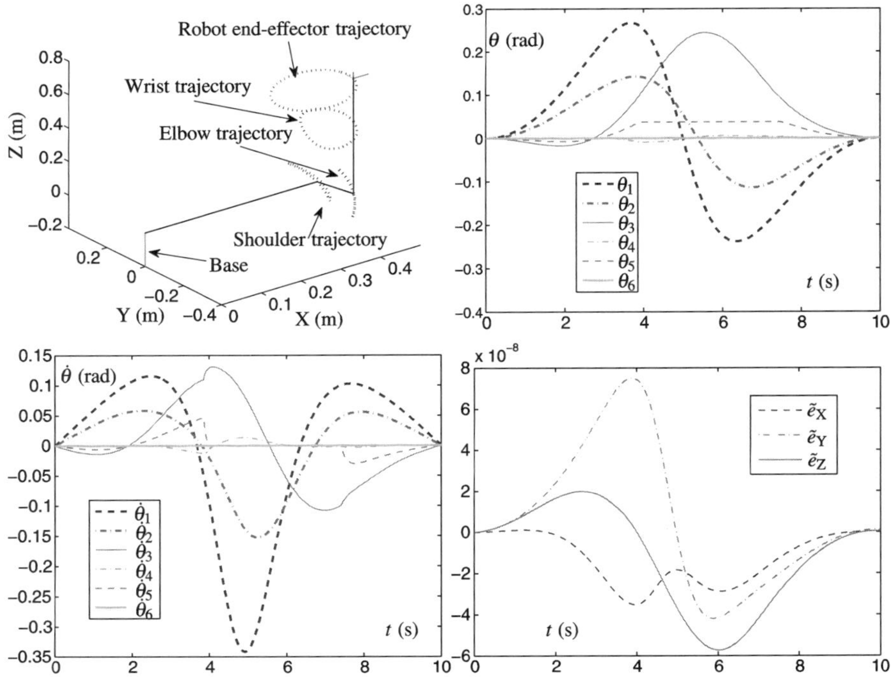

Fig. 9.5 PUMA560 motion trajectories and transients when its end-effector moves along a circle, which is synthesized by the simplified LVI-PDNN

9.2.2 Straight-Line Path Tracking

In this subsection, the PUMA560 end-effector is expected to move forwards and then backwards along a straight-line segment path. The straight-line segment with length being 2.25 m, at every motion cycle, starts with the PAUMA560 initial state $\theta(0) = [0, 0, 0, 0, 0, 0]^{\mathrm{T}}$ rad and will finally return to the initial state. Angles of the desired straight line making with XY, YZ, and ZX planes are respectively 0 rad, $\pi/4$ rad, and $\pi/4$ rad. The task duration at every motion cycle is 20 s.

The drift-free criterion and the simplified LVI-PDNN are applied to the control of such a physically constrained PUMA560 robot manipulator. In this straight-line path tracking example, the design parameters of the simplified LVI-PDNN are selected to be the same as those in the circular-path tracking example. As shown in Fig. 9.6, the final state of the PUMA560 manipulator coincides with its initial state very well. The final values of joint-angle variables are given in Table 9.3. In addition, all joint variables remain within their limited physical ranges.

Similar to the preceding circular-path tracking example, this straight-line path tracking example also demonstrates the efficacy of the drift-free scheme formulation and its online S-LVI-PDNN solution on the real-time redundancy resolution of physically constrained robot manipulators such as the PUMA560 manipulator.

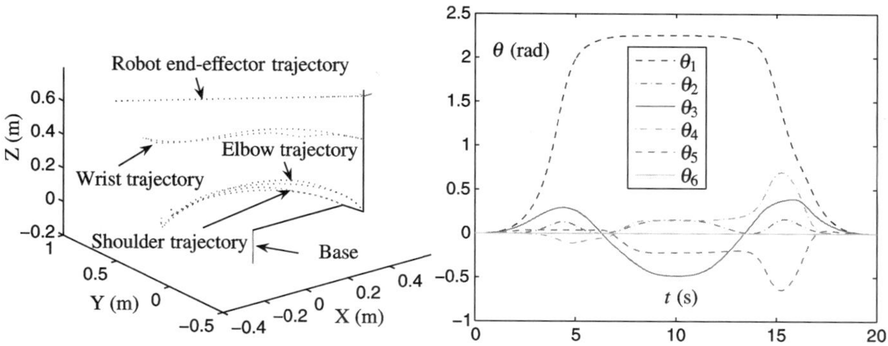

Fig. 9.6 PUMA560 motion trajectories and transients when its end-effector moves forwards and backwards along a straight-line segment path

Table 9.3 Final values of joint variables of PUMA560 robot manipulator after tracking a straight-line segment with drift-free criterion considered

#	joint	$\theta(0)$ (rad)	$\theta(20)$ (rad)
1	θ_1	0	$-0.00024998692356 \times 10^{-5}$
2	θ_2	0	$+0.00023455443849 \times 10^{-5}$
3	θ_3	0	$-0.00659146753014 \times 10^{-5}$
4	θ_4	0	$-0.16540394575433 \times 10^{-5}$
5	θ_5	0	$+0.01627381785611 \times 10^{-5}$
6	θ_6	0	$0.00000000000000 \times 10^{-5}$

In summary, as illustrated via the PUMA560 robot manipulator, we have numerically solved the joint-angle drift problem of physically constrained redundant manipulators. The solution is based on a quadratic-programming problem formulation and a simplified LVI-PDNN as the real-time QP solver. Simulation results based on the PUMA560 robot manipulator have further substantiated the efficacy of the proposed neural drift-free motion planning of joint-constrained redundant manipulators.

9.3 Chapter Conclusions

In this chapter, the physically constrained RMP scheme has been applied to the PUMA560 robot manipulator for the purpose of repetitive motion planning by using the LVI-PDNN and the simplified LVI-PDNN. Computer-simulation results with two different kinds of end-effector paths have further demonstrated the efficacy of the RMP scheme and the QP solvers (i.e., the LVI-PDNN and the simplified LVI-PDNN) on drift-free redundancy resolution of robot manipulator.

References

1. Zhang Y, Lv X, Li Z, Yang Z (2007) Repetitive motion planning of redundant robots based on LVI-based primal–dual neural network and PUMA560 example. In: Proceedings of life system modeling and simulation, pp 536–545
2. Zhang Y, Zhu H, Lv X, Li K (2008) Joint angle drift problem of PUMA560 robot arm solved by a simplified LVI-based primal–dual neural network. In: Proceedings of IEEE international conference on industrial technology, pp 1–6

Chapter 10
PA10 Examples

Abstract In this chapter, the physically constrained RMP scheme is applied to the PA10 robot manipulator by using two QP solvers, i.e., the LVI-PDNN and DNN. Different desired motion paths of the PA10 end-effector (i.e., circular and straight-line paths) are tested for illustrative and comparative purposes.

10.1 Simulations Using LVI-PDNN Solver

In this section, the physically constrained RMP scheme is solved by an LVI-PDNN QP solver and simulated based on the PA10 robot manipulator (which has seven degrees of freedom) [1]. The mechanical configuration and coordinate system, together with other specifications, of the PA10 redundant manipulator can be found in Chap. 1 of this book and also in [2, 3]. Note that the joint physical limits (i.e., joint-angle limits $\theta^{\pm}$ and joint-velocity limits $\dot{\theta}^{\pm}$) of the PA10 robot manipulator used in the simulations are given in Table 10.1. In addition, the PA10 end-effector is expected to track the circular and straight-line paths via the presented RMP scheme (2.2)–(2.5) and LVI-PDNN (5.36). The corresponding simulation results are shown as follows.

10.1.1 Circular Path Tracking

In simulations of this subsection, the motion trajectory of the PA10 end-effector is expected to be a circular path with radius being 0.2 m and the revolute angle about X-axis being $\pi/6$ rad. The task duration is 10 s, and the initial joint state $\theta(0) = [0, \pi/4, 0, 2\pi/3, 0, -\pi/4, 0]^{\mathrm{T}}$ rad.

First, we show, in Fig. 10.1, the redundancy-resolution results with neither joint physical limits nor drift-free criterion considered. That is, $\theta^{\pm}$ and $\dot{\theta}^{\pm}$ can be set as $\pm\varpi$, while the drift-free coefficient $\lambda = 0$. As seen from the left graph of Fig. 10.1, the end-effector of the PA10 manipulator moves along a circular path in the three-dimensional workspace, which is sufficiently close to the desired circular path. However, as seen from the right graph of Fig. 10.1, the joint angle θ_4 exceeds its limited range $[-2.6831, 2.6831]$ rad. In addition, such a solution is not repetitive

Y. Zhang, Z. Zhang, *Repetitive Motion Planning and Control of Redundant Robot Manipulators*, DOI 10.1007/978-3-642-37518-7_10,

Table 10.1 Joints description and physical limits of PA10 robot manipulator

#	joint	axis	$\theta^{\pm}$	$\dot{\theta}^{\pm}$
1	shoulder 1	rotating	$\pm\pi$ rad	± 1 rad/s
2	shoulder 2	pivot	± 1.7637 rad	± 1 rad/s
3	shoulder 3	rotating	$\pm\pi$ rad	± 2 rad/s
4	elbow 1	pivot	± 2.6831 rad	± 2 rad/s
5	elbow 2	rotating	$\pm 3\pi/2$ rad	$\pm 2\pi$ rad/s
6	wrist 1	pivot	$\pm\pi$ rad	$\pm 2\pi$ rad/s
7	wrist 2	rotating	$\pm 2\pi$ rad	$\pm 2\pi$ rad/s

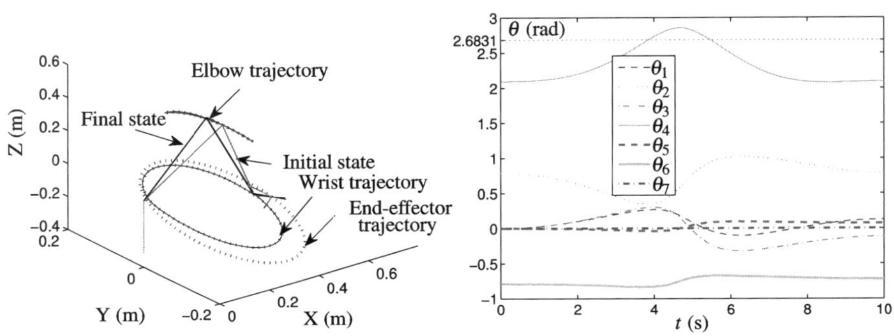

Fig. 10.1 PA10 end-effector moves along a circular path with neither joint physical limits nor drift-free criterion considered

because $\theta_1(0) \neq \theta_1(10)$, $\theta_3(0) \neq \theta_3(10)$, and $\theta_5(0) \neq \theta_5(10)$. Thus, in the context of constrained repetitive motion planning, such a solution is practically inapplicable. On one hand, if it is exploited directly to control the PA10 manipulator with θ_4 finally locked at value 2.6831 rad, then the tracking error may increase considerably, in addition to the physical damage probably caused. On the other hand, if such a nonrepetitive solution is exploited to control the PA10 manipulator, then an additional self-motion readjustment may be needed, which would be less efficient and less desirable in engineering applications.

Second, for comparison, we show, in Fig. 10.2, the redundancy-resolution results with joint physical limits considered but without considering drift-free criterion. That is, $\theta^{\pm}$ and $\dot{\theta}^{\pm}$ are set correctly as in Table 10.1, while the drift-free coefficient λ is still zero. As seen from the left graph of Fig. 10.2, the end-effector of the PA10 manipulator moves along a circular path in the three-dimensional workspace accurately as well. In addition, as seen from the right graph of Fig. 10.2 and Table 10.1, all joint angles are kept within their limited ranges. However, for robots working with a cyclic motion requirement, such a solution has a serious weakness. That is, the final state of the solution does not coincide with its initial state; mathematically, $\theta_1(0) \neq \theta_1(10)$, $\theta_3(0) \neq \theta_3(10)$, $\theta_4(0) \neq \theta_4(10)$, $\theta_5(0) \neq \theta_5(10)$, and $\theta_6(0) \neq \theta_6(10)$.

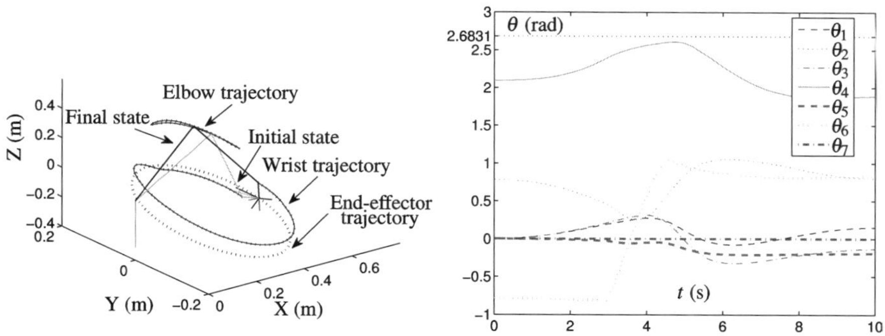

Fig. 10.2 PA10 end-effector moves along a circular path with joint physical limits considered but without considering drift-free criterion

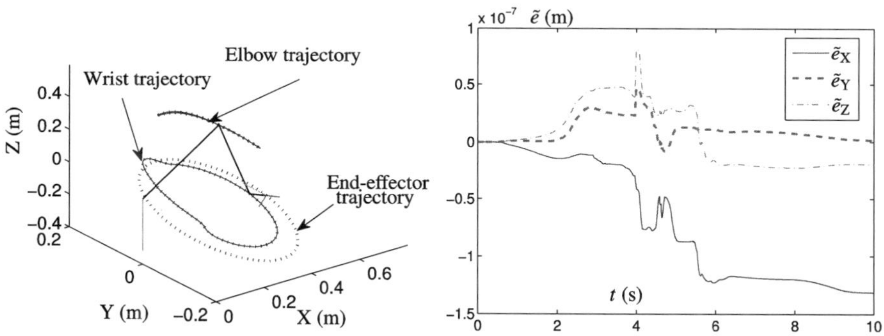

Fig. 10.3 PA10 end-effector moves along a circular path with both joint physical limits and drift-free criterion considered, and PA10 final state coincides with the initial state

Third, for comparison and illustration, we show, in Figs. 10.3 and 10.4, the redundancy-resolution results with both joint physical limits and drift-free criterion considered. That is, $\theta^{\pm}$ and $\dot{\theta}^{\pm}$ are set correctly, and $\mu = 1$ and $\lambda = 4$ are set. As seen from the left graph of Fig. 10.3, the end-effector of the PA10 manipulator moves along a circular path in the three-dimensional workspace, which is sufficiently close to the desired circular path with a tiny position error (i.e., less than 1.5×10^{-7} m as shown in the right graph of Fig. 10.3). In addition, as seen from Table 10.1 and Fig. 10.4, all joint-angle and joint-velocity variables are kept within their mechanical ranges. Moreover, this solution (including θ and $\dot{\theta}$) is repetitive and applicable for PA10 working with a cyclic motion requirement because the final and initial states of the PA10 manipulator coincide with each other. It is worth noting that, in the right graph of Fig. 10.3, we use $\tilde{e}_{\text{X}}$, $\tilde{e}_{\text{Y}}$, and $\tilde{e}_{\text{Z}}$ again to denote respectively the X-, Y-, and Z-axis components of the position error $\tilde{e}(t) := r(t) - f(\theta(t))$ with respect to the base frame and that, in the lower left graph of Fig. 10.4, we use $\dot{\tilde{e}}_{\text{X}}$, $\dot{\tilde{e}}_{\text{Y}}$, and $\dot{\tilde{e}}_{\text{Z}}$ again to denote the corresponding components of velocity error

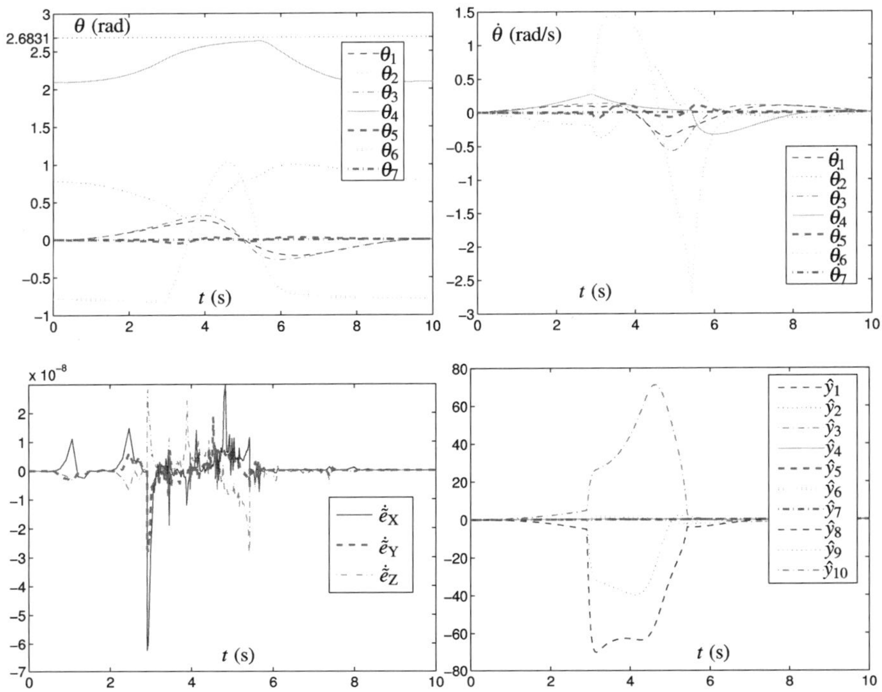

Fig. 10.4 PA10 joint transients and end-effector velocity error as well as LVI-PDNN state transients, which correspond to Fig. 10.3

$\dot{e}(t) := \dot{r}(t) - J(\theta(t))\dot{\theta}(t)$, with the maximal velocity error less than 7.0×10^{-8} m/s in the graph.

In summary, the above circular-path tracking example has demonstrated the capability of the physically constrained RMP scheme and the corresponding LVI-PDNN QP solver for the online repetitive motion planning of physically constrained robot manipulators.

10.1.2 Straight-Line Path Tracking

In simulations of this subsection, the motion trajectory of the PA10 end-effector is expected to move forward and then return backward along a straight-line path. The duration of this path-tracking task is 7 s. In addition, the length of such a straight-line path is 2.5 m, and the robot is required to start from initial state $\theta(0) = [\pi/4, \pi/4, \pi/4, \pi/4, \pi/4, \pi/4, 0]^{\mathrm{T}}$ rad and finally return to the initial state. Angles of the desired straight-line making with XY, YZ, and ZX planes are respectively $\pi/4$ rad, $\pi/6$ rad, and $\pi/6$ rad. For comparisons, we perform the following four simulation tests.

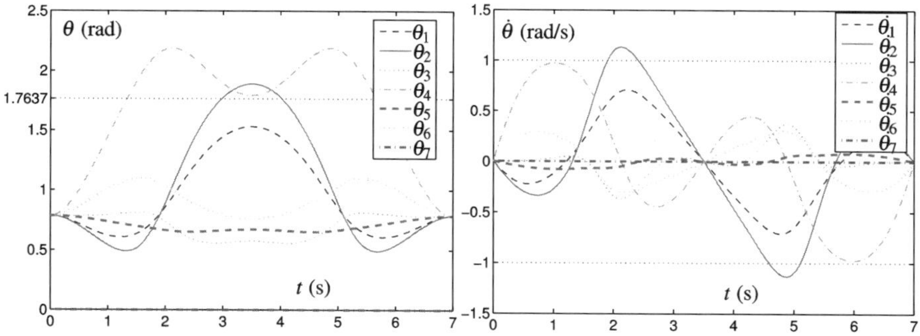

Fig. 10.5 PA10 end-effector moves along a straight-line segment with neither joint physical limits nor drift-free criterion considered

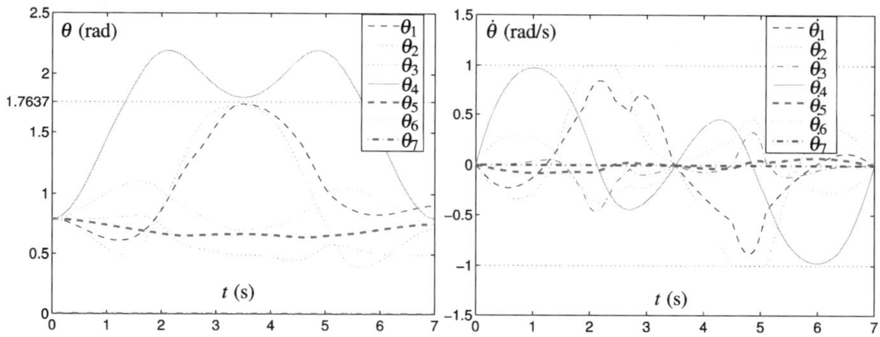

Fig. 10.6 PA10 end-effector moves along a straight-line segment with joint physical limits considered but without considering drift-free criterion

First, with neither joint physical limits nor drift-free criterion considered, the inverse-kinematics problem of the PA10 manipulator is solved via LVI-PDNN (5.36). The result is shown in Fig. 10.5. It is seen from the figure that the solution happens to be repetitive but is inapplicable because the joint angle $\theta_2(t)$ hits its upper limit $\theta_2^+ = 1.7637$ rad and its velocity $\dot{\theta}_2(t)$ also exceeds its physical limits by ± 1 rad/s.

Second, with joint physical limits considered (i.e., $\mu = 4$) but without considering drift-free criterion, the inverse-kinematics problem is solved again via LVI-PDNN. The result is shown in Fig. 10.6, wherein all joint angles and joint velocities are kept within their mechanical ranges. However, this solution is still not applicable in the context of cyclic motion planning because the joint-drift phenomenon occurs. That is, $\theta_i(0) \neq \theta_i(7)$ for $i = 1, 2, 3, 5, 6$. Besides, by comparing this simulation with the preceding one, it is worth noting that avoiding joint physical limits may unexpectedly induce a joint-drift problem, which, as a fact, may be explained as follows. On one hand, if there is no joint physical limit in the PA10 robot manipulator, for a closed end-effector path (such as the aforementioned straight-line segment), there may exist closed trajectories in the joint space. On the other hand,

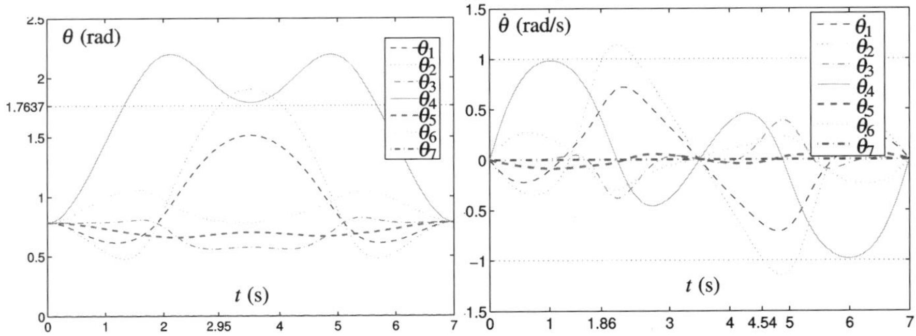

Fig. 10.7 PA10 end-effector moves along a straight-line segment with drift-free criterion considered but without considering joint physical limits

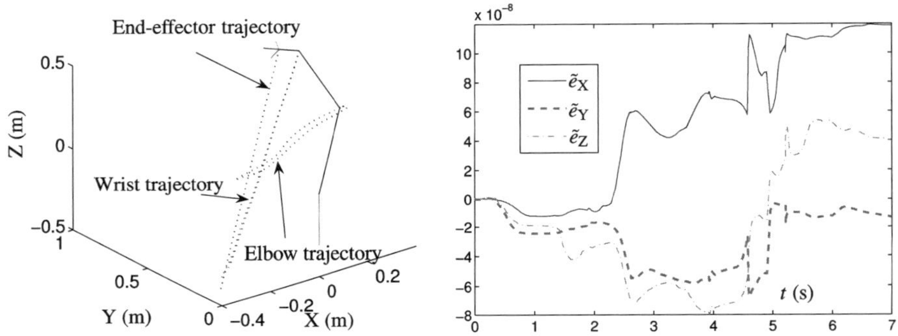

Fig. 10.8 PA10 end-effector moves along a straight-line segment with both joint physical limits and drift-free criterion considered, and PA10 final state coincides with the initial state

as joint physical limits do exist for almost any robot, people have to consider the avoidance of joint physical limits in robot control, which may thus disconnect those closed joint-trajectories with the joint-drift phenomenon occurring.

Third, with drift-free criterion considered (i.e., $\lambda = 4$) but without considering joint physical limits, the PA10 inverse-kinematics problem is solved, and the result is shown in Fig. 10.7. Simply speaking, though repetitive, the solution is inapplicable again due to the violation of joint physical limits.

Fourth, for comparison and illustration, the PA10 inverse-kinematics problem is finally solved via LVI-PDNN with both joint physical limits and drift-free criterion considered. The results are shown in Figs. 10.8 and 10.9. As seen from the left graph of Fig. 10.8, the end-effector of the PA10 manipulator moves along a straight-line in the three-dimensional workspace, which is sufficiently close to the desired path with a tiny position error (i.e., less than 1.2×10^{-7} m as shown in the right graph of Fig. 10.8). In addition, as seen from Table 10.1 and Fig. 10.9, all joint-angle and joint-velocity variables are kept within their mechanical ranges. Moreover, this solution (including θ and $\dot{\theta}$) is repetitive and applicable for PA10 working with a

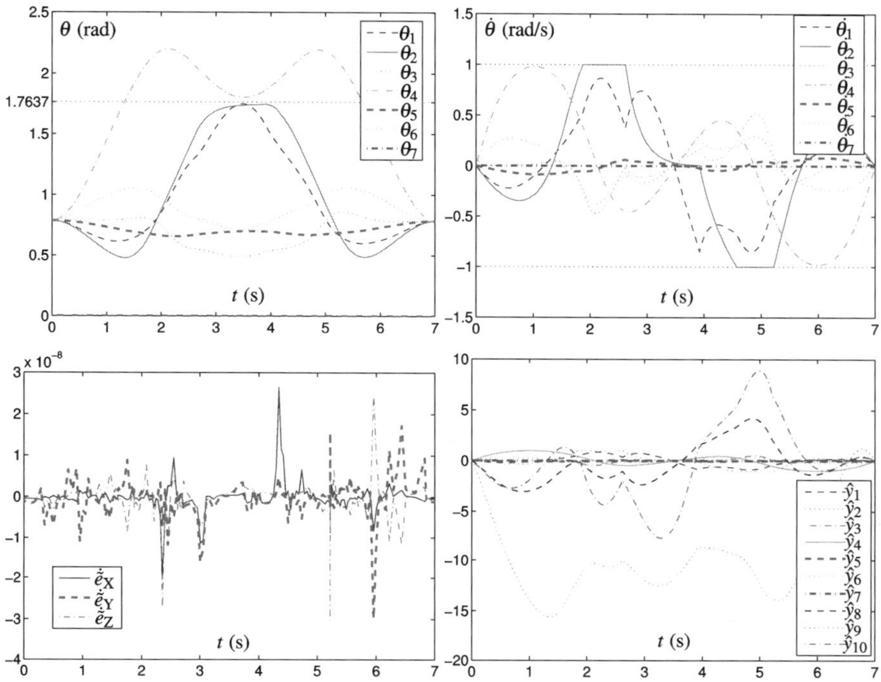

Fig. 10.9 PA10 joint transients and end-effector velocity error as well as LVI-PDNN state transients, which correspond to Fig. 10.8

cyclic-motion requirement because the final and initial states of the PA10 manipulator coincide. Note that, as shown in the lower left graph of Fig. 10.9, the maximal velocity error is about 3×10^{-8} m/s.

In summary, similar to the preceding circular-path tracking example, this straight-line tracking example has demonstrated the capability of the LVI-PDNN-based redundancy-resolution scheme for the online drift-free motion planning of physically constrained robot manipulators.

10.2 Simulations Using DNN Solver

In this section, we apply the DNN to the RMP redundancy resolution of the physically constrained PA10 manipulator [4]. Simulations are performed for the path-tracking task that the end-effector of the PA10 manipulator moves along a given circle or straight-line segment in the three-dimensional workspace. The joint limits $\theta^{\pm}$ and joint-velocity limits $\dot{\theta}^{\pm}$ are shown in Table 10.1.

It is worth pointing out that the limited joint range $[\theta^-, \theta^+]$ in this section is converted into the following joint-velocity bound constraint:

$$\mu_1\big(\mu_2\theta^- - \theta\big) \le \dot{\theta} \le \mu_1\big(\mu_2\theta^+ - \theta\big), \tag{10.1}$$

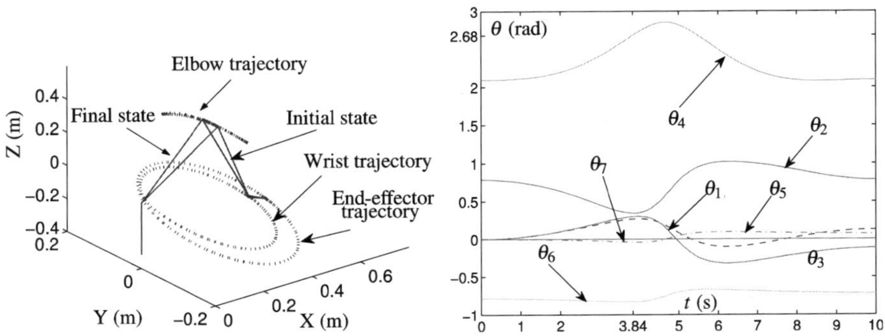

Fig. 10.10 End-effector of PA10 manipulator moves along a circular path in the three-dimensional workspace with neither joint physical limits nor drift-free criterion considered

where coefficient $\mu_2 \in (0, 1]$ defines the critical region $(\theta^-, \mu_2\theta^-]$ or $[\mu_2\theta^+, \theta^+)$ as a buffer for the situation of $\theta^- < 0 < \theta^+$ to avoid hitting the joint limits, and $\mu_1 > 0$ should be large enough to scale the feasible region of $\dot{\theta}$.

10.2.1 Circular Path Tracking

In simulations of this subsection, we first solve the redundancy resolution problem without considering joint physical constraints and drift-free criterion. Specifically, the proposed DNN with $\theta^\pm$ and $\dot{\theta}^\pm$ disabled and the drift-free coefficient $\lambda = 0$ is applied to the PA10 robot manipulator.

The desired motion of the PA10 end-effector is a circular path with radius 0.2 m and revolute angle about X-axis being $\pi/6$ rad. The task duration of the motion is 10 s, and the initial joint state $\theta(0) = [0, \pi/4, 0, 2\pi/3, 0, -\pi/4, 0]^\mathrm{T}$ rad. Figure 10.10 illustrates the motion trajectories of the PA10 manipulator when its end-effector moves along a circular path in the three-dimensional workspace and correspondingly the transients of joint angles. Although the maximal Cartesian position and velocity tracking errors are less than 1.5×10^{-7} m and 6.0×10^{-8} m/s, respectively, the joint angle θ_4 exceeds its mechanical range $[-2.6831, 2.6831]$, and thus the solution is actually inapplicable. If such a solution is directly applied to the PA10 manipulator with θ_4 finally locked at value -2.6831 rad, the tracking error increases considerably, in addition to the physical damage possibly caused. Hence, the path-tracking task fails. Moreover, as seen form Fig. 10.10, the solution is not repetitive in the sense that the final state of the PA10 manipulator does not coincide with its initial state, i.e., $\theta_1(10) \neq \theta_1(0)$, $\theta_3(10) \neq \theta_3(0)$, and $\theta_5(10) \neq \theta_5(0)$. Hence, if we accept the solution, an inefficient and undesirable readjustment is needed for the cyclic motion control. In summary, the above simulation results show that physical constraints and drift-free criterion are in general worth considering in the repetitive path-tracking tasks.

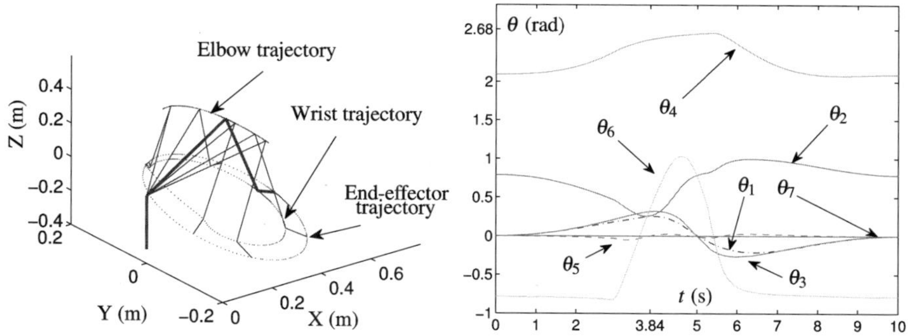

Fig. 10.11 End-effector of PA10 manipulator moves along a circular path in the three-dimensional workspace with both joint physical limits and drift-free criterion considered

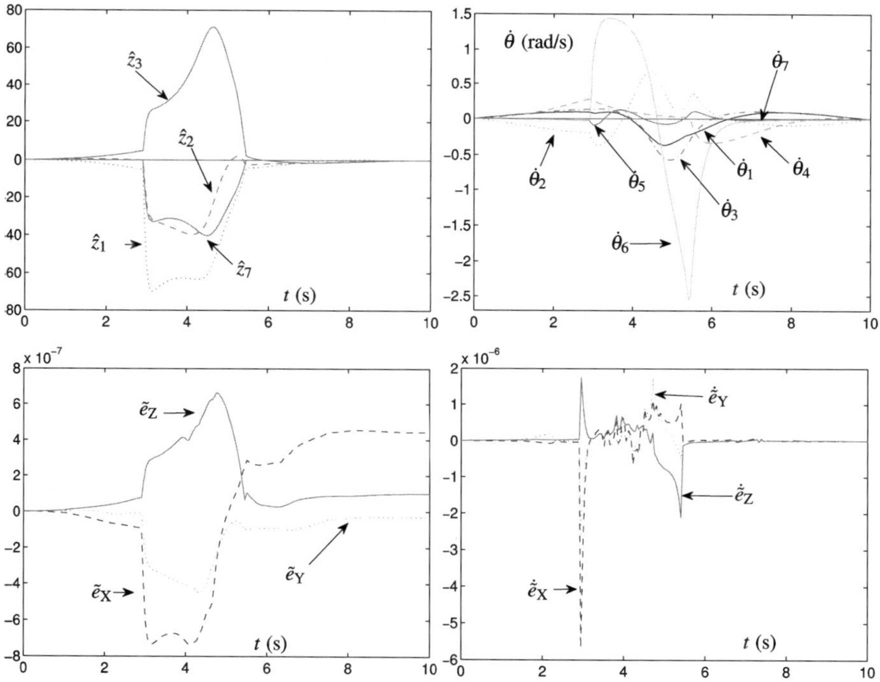

Fig. 10.12 Transients of the DNN and the PA10 manipulator moving along a circular path

Now let us consider the joint limits in Table 10.1 and the drift-free criterion as well. Specifically, the DNN, with $\mu_2 = 0.99$ and $\mu_1 = 1$ in (10.1) and $\lambda = 4$ in (2.2), is applied to PA10 for the same circular-path tracking task. Figure 10.11 shows the three-dimensional motion trajectories of the PA10 manipulator and the corresponding joint angles. Figure 10.12 illustrates the transient behaviors of the DNN and the PA10 manipulator, including the profiles of dual decision vector $\hat{z}$, the PA10

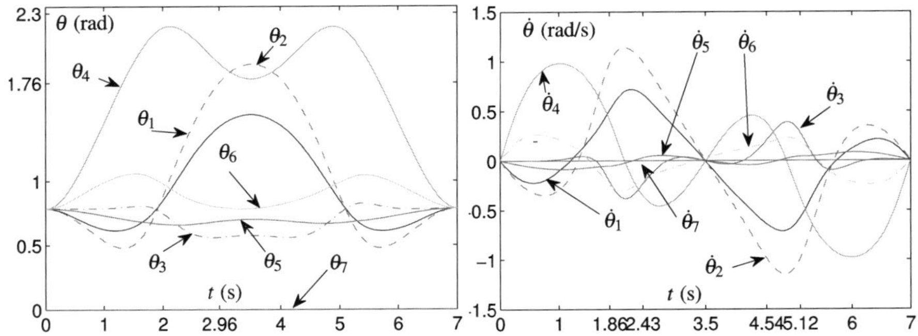

Fig. 10.13 End-effector of PA10 manipulator moves along a straight-line segment with drift-free criterion considered but without considering joint physical limits

joint velocity $\dot{\theta}$, Cartesian position error $\tilde{e}$, and velocity error $\dot{\tilde{e}}$. More specifically, as seen in Fig. 10.11, the joint angle θ_4 now never exceeds the mechanical range $[-2.6831, 2.6831]$, and the solution is repetitive in the sense that the initial state and final state of the PA10 manipulator coincide with each other. It follows from Fig. 10.12 that no joint-velocity variable exceeds its limits in Table 10.1. In addition, the maximal position and velocity errors are less than 8×10^{-7} m and 6×10^{-6} m/s, respectively.

In summary, the above circular-path tracking simulations have demonstrated the capability of the DNN for resolving online the drift-free redundancy of physically constrained manipulators.

10.2.2 Straight-Line Path Tracking

In simulations of this subsection, the PA10 manipulator is controlled to move forward and then return backward along a straight-line segment. The length of the straight-line segment is 2.5 m. At every motion cycle, the PA10 robot starts from the initial state $\theta(0) = [\pi/4, \pi/4, \pi/4, \pi/4, \pi/4, \pi/4, 0]^{\mathrm{T}}$ rad and will finally return to the initial state. The angles of the desired straight-line segment made with XY, YZ, and ZX planes are, respectively, $\pi/4$ rad, $\pi/6$ rad, and $\pi/6$ rad. The duration of the path-tracking task at every motion cycle is specified as 7.0 s.

For comparison, the drift-free inverse kinematics problem is first solved without considering joint physical constraints, as shown in Fig. 10.13. Clearly, the solution is not acceptable in practice since θ_2 hits its mechanical limit $\theta_2^+ = 1.7637$ rad at time $t = 2.95$ s and $\dot{\theta}_2$ also exceeds the joint-velocity limits at $t = 1.86$ s and at $t = 4.54$ s.

The DNN is then applied to the physically constrained PA10 manipulator for the drift-free redundancy resolution. Parameters of the dual neural network are selected the same as in the circular example except $\mu_1 = 2$. Simulation results are depicted in Figs. 10.14 and 10.15. The transient behaviors of joint-angle and joint-velocity

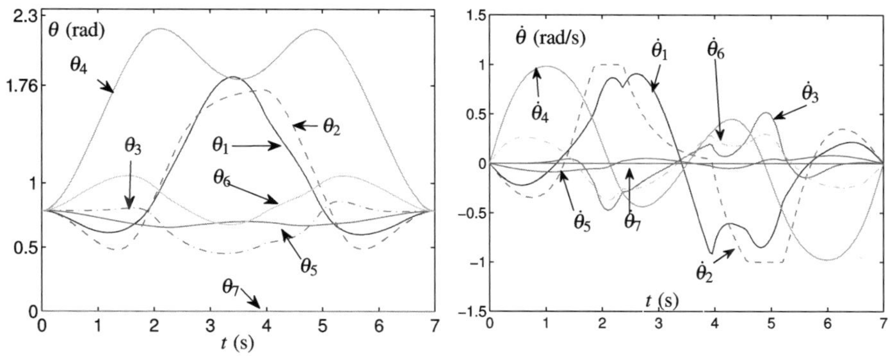

Fig. 10.14 End-effector of PA10 manipulator moves along a straight-line segment with both joint physical limits and drift-free criterion considered

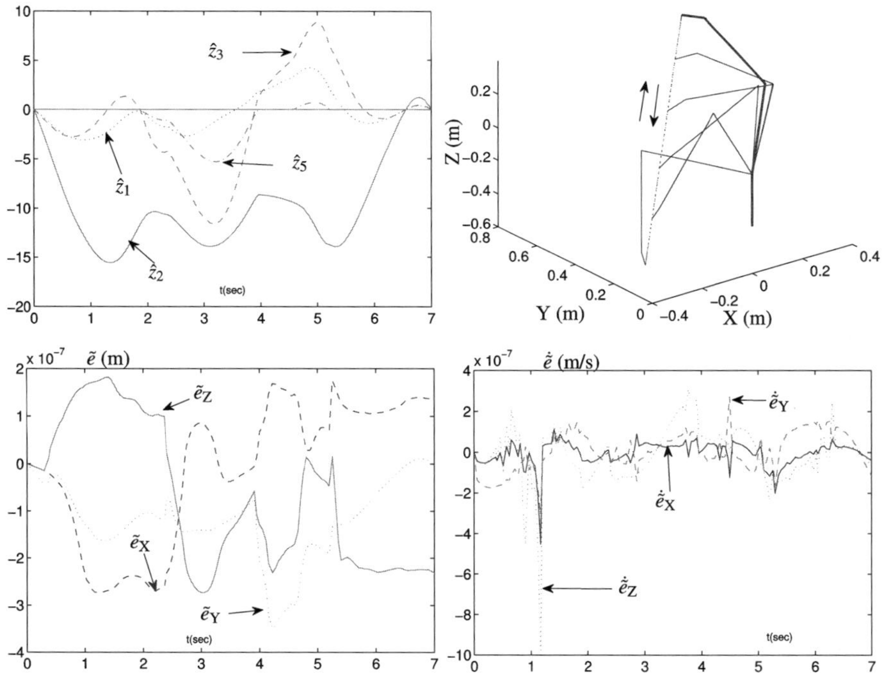

Fig. 10.15 Transients of the DNN and the PA10 manipulator moving along a straight-line segment

variables are depicted in Fig. 10.14. Compared with the left graph of Fig. 10.13, the joint variable θ_2 in the left graph of Fig. 10.14 does not exceed its upper limit 1.7637 rad, and all other joint angles remain in their limited ranges. Compared with the right graph of Fig. 10.13, the joint velocity $\dot{\theta}(t)$ including $\dot{\theta}_2$, as shown in the right graph of Fig. 10.14, is kept within the mechanical range. The maximal position and velocity errors are respectively less than 4×10^{-7} m and 10^{-6} m/s, which are shown in Fig. 10.15.

In summary, the above simulation results have substantiated the efficacy of the DNN approach to online drift-free (or to say, RMP) redundancy resolution for physically constrained manipulators.

10.3 Chapter Conclusions

In this chapter, the physically constrained RMP scheme has been applied to the PA10 robot manipulator for the purpose of repetitive motion planning by using the LVI-PDNN and DNN solvers. Computer-simulation results with different end-effector trajectories have further demonstrated the efficacy of the QP-based RMP scheme and of the QP solvers (i.e., the LVI-PDNN and DNN) on the drift-free redundancy resolution of robot manipulator.

References

1. Zhang Y, Lv X, Li Z, Yang Z, Chen K (2008) Repetitive motion planning of PA10 robot arm subject to joint physical limits and using LVI-based primal–dual neural network. Mechatronics 18(9):475–485
2. Wang J, Zhang Y (2004) Recurrent neural networks for real-time computation of inverse kinematics of redundant manipulators. In: Sincak P, Vascak J, Hirota K (eds) Machine intelligence: quo vadis? World Scientific, Singapore, pp 299–319
3. Wang J, Hu Q, Jiang D (1999) A Lagrangian network for kinematic control of redundant manipulators. IEEE Trans Neural Netw 10(5):1123–1132
4. Zhang Y, Wang J, Xia Y (2003) A dual neural network for redundancy resolution of kinematically redundant manipulators subject to joint limits and joint velocity limits. IEEE Trans Neural Netw 14(3):658–667

Chapter 11
Physical Robot Manipulator Experiments

Abstract To demonstrate the hardware realizability and efficacy of the QP-based method for solving the nonrepetitive problem, this chapter realizes the physically constrained scheme on a physical planar six-degrees-of-freedom (DOF) push-rod-joint (PRJ) manipulator. To control the PRJ manipulator, this scheme considers variable joint-velocity limits and joint-limit margins, and to decrease the model disturbance and computational round-off errors, this scheme also considers the position-error feedback. Then, the scheme is reformulated as a QP problem. Due to the control performed on a digital computer, the discrete-time QP solver, i.e., the 94LVI numerical algorithm, is used to solve the QP problem. For comparison, both of the nonrepetitive and repetitive motions are performed on the six-DOF PRJ manipulator to track square, B-shaped, and circular paths. Theoretical analysis and experimental results validate the physical realizability and effectiveness of the physically constrained RMP scheme. Position-error analysis further verifies the accuracy of this scheme.

11.1 Introduction

Repetitive motion planning (RMP) of a redundant manipulator is an appealing topic in the field of robotics because the nonrepetitive phenomenon of the manipulator may induce a problem that the behavior of the manipulator is hard to predict and readjusting the configuration of the manipulator through self-motion is less efficient [1, 2]. In the past decades, various schemes have been proposed and investigated to solve such a nonrepetitive problem [3–5]. Specifically, Shamir and Yomdin [3] explained why the nonrepeatability occurred for redundant robot manipulators. De Luca et al. [4] presented a Jacobian pseudoinversion-based resolution method that could achieve asymptotic cyclicity. Bowling and Harmeyer [5] used a nullspace quasi-velocities approach to remedy the nonrepetitive problem. In essence, the above studies are based on the pseudoinverse methods. Recent research shows that the solution to nonrepetitive motion problems can be enhanced using optimization techniques. Furthermore, most of these optimization methods can be converted into QP problems that are solved readily by QP solvers [6, 7]. Specifically, presented in [6] and Chap. 5 of this book, the LVI-based primal–dual neural network can be used to solve the resultant QP problem and the RMP scheme. Presented in [7]

Y. Zhang, Z. Zhang, *Repetitive Motion Planning and Control of Redundant Robot Manipulators*, DOI 10.1007/978-3-642-37518-7_11,

and Chaps. 4 and 5, three kinds of recurrent neural networks can be used to solve the QP problem so as to achieve repetitive motion. In addition, in Chaps. 8 through 10, various QP solvers and robots are presented to demonstrate the effectiveness of the RMP scheme based on computer simulations. These results demonstrate that such QP-based methods are effective. However, most of these existing studies of such QP-based RMP schemes are only conducted and verified on computer simulations. Therefore, it is necessary to investigate and discuss the hardware realizability and effectiveness via the physical robots. Besides, the joint-velocity limits of the existing studies are considered to be constant, which may not be applicable to the push-rod-joint (PRJ) manipulators. Furthermore, among the published papers, the QP solvers adopt mainly continuous-time models (e.g., the neural networks), of which the efficacy and superiority are based on the specialized circuit implementation of the models. Note that such continuous-time models are not applicable easily to the experimental verification via digital computers and that many existing QP-based RMP schemes do not consider the feedback control, which may be less appropriate in engineering applications.

This chapter focuses on the experimental implementation and verification of the novel physically constrained RMP scheme on an actual redundant robot manipulator. To avoid some joint(s) being locked by safety devices in practical applications, the new bound constraint with sufficient joint-limit margins is designed and exploited. Note that the joint-velocity limits in the proposed physically constrained scheme vary with the end-effector and joint motion, which is considered and investigated for the first time. In addition, unlike the aforementioned usual continuous-time QP solvers, this chapter exploits and investigates a discrete-time QP solver, i.e., the 94LVI numerical algorithm presented in Chap. 6, which is readily realized on the digital computer and the physical robot system. Furthermore, the position-error feedback is considered in the presented physically constrained scheme, which can decrease effectively the model disturbance and computational round-off errors.

11.2 Robot-Manipulator Preliminaries

The general redundancy resolution process of such a physically constrained scheme is divided into four steps. First, the desired path r is given and differentiated with respect to time t. Then, the resultant $\dot{r}$ is incorporated into the proposed RMP scheme [e.g., (2.2)–(2.5)]. Second, the scheme is reformulated as a standard QP problem. Third, a QP solver (e.g., a numerical algorithm) is used to solve the resultant QP problem. Finally, the computed data (e.g., values of joint angle θ and joint velocity $\dot{\theta}$) are used to control the manipulator to perform the desired path-tracking task.

As mentioned previously, the computer simulations and experiments on RMP in this chapter are performed on the six-DOF PRJ redundant robot manipulator which is shown in Fig. 11.1. The left photo of Fig. 11.1 shows the manipulator in the laboratory, and its corresponding 3-D model is presented in the right image. Furthermore, as shown in the left photo of Fig. 11.2, Joints 2–6 of the robot are "shaped" as

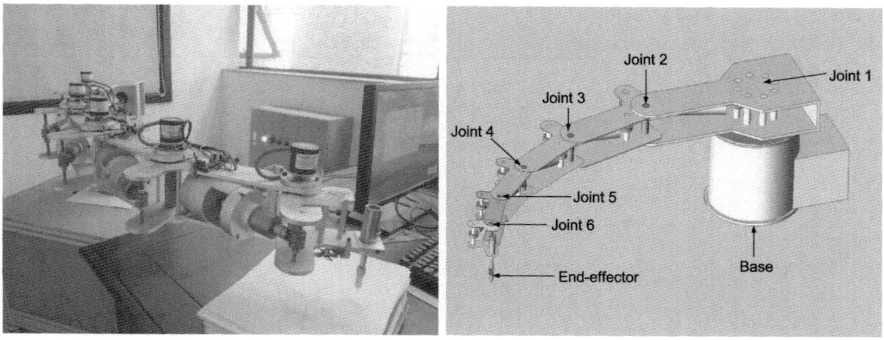

Fig. 11.1 Hardware system and 3-D model of the six-DOF PRJ manipulator

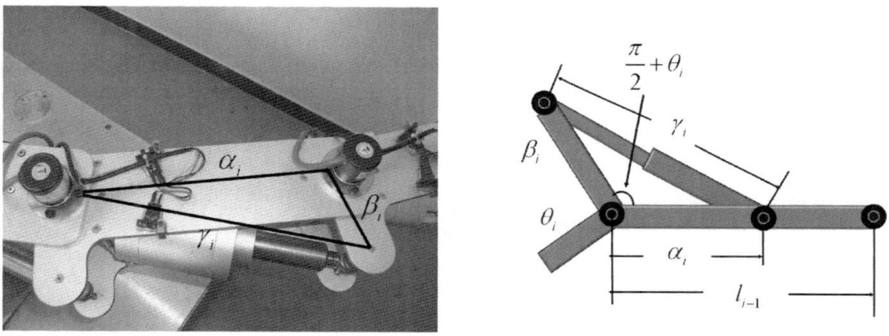

Fig. 11.2 Local configuration of the six-DOF PRJ manipulator

triangles. The three edges of the ith triangle are α_i, β_i, and γ_i (with $i = 2, 3, \ldots, 6$) shown on the right image of Fig. 11.2. The physical parameters of the joints are shown in Table 11.1, wherein l_i denotes the length of the ith link, $i = 1, 2, \ldots, 6$.

The whole robot-arm system is composed of a host computer and a manipulator. The host computer is a personal digital computer with a Pentium E5300 2.6 GHz CPU, 4 GB DDR3 memory, and a Windows XP Professional operating system, sending instructions to the manipulator motion control module with six-axis control card of peripheral component interconnect (PCI). Stepper motors are employed in the other five joints, except for the first joint using a servo motor. According to the instructions, the motion-control module of the manipulator converts the data into the actual pulse signals to drive the motors (i.e., the servo and stepper motors).

11.3 Scheme Formulation and QP Solver

Generally, the joint velocity $\dot{\theta}$ is first computed via the QP-based scheme and the QP solver at each sampling interval according to the desired end-effector velocity $\dot{r}$. This is the motion planning process. Then, the motors of the joints are driven by the

Table 11.1 Physical joint parameters of six-DOF PRJ redundant robot manipulator

#	θ_i^- (rad)	θ_i^+ (rad)	l_i (m)	α_i (m)	β_i (m)
$i=1$	−1.536	1.431	0.301	–	–
$i=2$	0.052	0.785	0.290	0.250	0.080
$i=3$	0.026	0.611	0.230	0.250	0.080
$i=4$	0.066	0.576	0.225	0.190	0.080
$i=5$	0.017	0.559	0.214	0.185	0.080
$i=6$	0.009	0.445	0.103	0.174	0.080

pulses transmitted from the host computer, i.e., the actual control input to the joints. This is the control process. Through the above two steps, the physically constrained RMP scheme for tracking a desired end-effector path can be achieved. It is worth pointing out that, to avoid some joint(s) being locked by safety devices in the experiments, a bound constraint with sufficient joint-limit margins should be designed and exploited based on the detailed analysis of the push-rod joints.

11.3.1 RMP for the Six-DOF PRJ Manipulator

As shown in Fig. 11.1, the first joint of the six-DOF PRJ manipulator is driven by a servo motor. The positive and negative rotation rate limits of the motor are ± 2500 rot/min, and the motor's reduction ratio is 80. Thus, the upper and lower limits of the first joint are $\dot{\theta}_1^{\pm} = \pm 2500 \times 2\pi/(80 \times 60) = \pm 25\pi/24$ rad/s, i.e., $-25\pi/24 \le \dot{\theta}_1 \le 25\pi/24$ rad/s.

For Joints 2 through 6 of the six-DOF PRJ manipulator, the variable joint-velocity limits (VJVL) $\dot{\theta}^{\pm}(\theta)$, which change with joint angle θ (i.e., being functions of θ), are obtained by following the law of cosines (with the joint-structure shown in Fig. 11.2). More specifically, with $i = 2, 3, \ldots, 6$,

$$\dot{\theta}_i^-(\theta_i) \le \dot{\theta}_i \le \dot{\theta}_i^{\,+}(\theta_i),$$

where

$$\dot{\theta}_i^{\pm}(\theta_i) = \frac{\tilde{s}_i \tilde{v}_i^{\pm} \sqrt{\alpha_i^2 + \beta_i^2 + 2\alpha_i\beta_i \sin\theta_i}}{\alpha_i\beta_i \cos\theta_i} \tag{11.1}$$

with $\tilde{v}_i^+$ and $\tilde{v}_i^-$ being the positive and negative rotation rate limits of the ith stepper motor and $\tilde{s}_i$ denoting the elongation rate of the ith push-rod (i.e., the elongation length when the motor moves a full turn). For this manipulator hardware system, $-\tilde{v}_i^- = \tilde{v}_i^+ = 10$ rot/s and $\tilde{s}_i = 2.5 \times 10^{-3}$ m/rot.

According to Eqs. (2.2)–(2.5) and VJVL (11.1) and considering the feedback of Cartesian position error $\tilde{e} = r - f(\theta)$ with $\hat{q} = \lambda(\theta(t) - \theta(0))$ and $\lambda > 0$ defined as

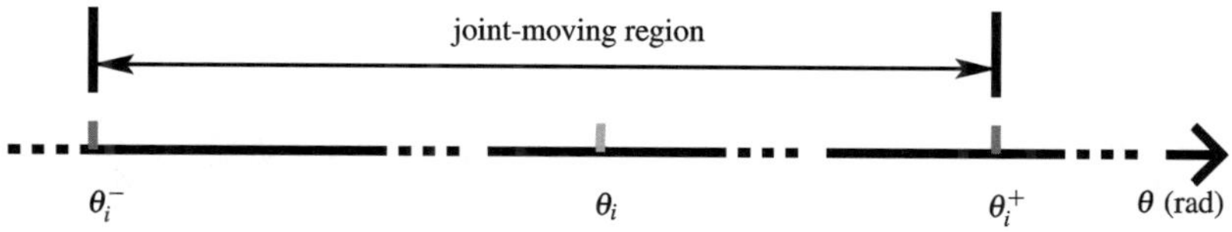

Fig. 11.3 Physical limits $\theta_i^{\pm}$ of the ith joint without considering joint-limit margins

before, the proposed scheme (2.2)–(2.5) can be reformulated as

$$\text{minimize} \quad (\dot{\theta} + \hat{q})^{\mathrm{T}}(\dot{\theta} + \hat{q})/2, \tag{11.2}$$

$$\text{subject to} \quad J(\theta)\dot{\theta} = \dot{r} + K\big(r - f(\theta)\big), \tag{11.3}$$

$$\theta^- \le \theta \le \theta^+, \tag{11.4}$$

$$\dot{\theta}^-(\theta) \le \dot{\theta} \le \dot{\theta}^+(\theta), \tag{11.5}$$

where K is a positive definite symmetric (typically diagonal) $m \times m$ feedback gain matrix (in the ensuing simulations and experiments, $K = 8I$ with I denoting the identity matrix). As noted in [8], applying $\dot{\tilde{e}} = -K\tilde{e}$ yields (11.3), which, as a feedback control method, can effectively decrease the accumulated errors of the end-effector positioning. In addition, $\dot{\theta}^+(\theta)$ and $\dot{\theta}^-(\theta)$ denote respectively the upper and lower limits of the joint-velocity vector $\dot{\theta}$, which are functions of the joint-angle vector θ. Other parameters have the same meanings as those in Eqs. (2.2)–(2.5).

11.3.2 Bound Constraint with Joint-Limit Margins

The joint physical limits (11.4) and (11.5) have to be transformed and incorporated into a bound constraint in terms of $\dot{\theta}$, i.e., $\xi^- \le \dot{\theta} \le \xi^+$ as the redundancy is resolved at the joint-velocity level. In [6, 7], the ith elements of ξ^- and ξ^+ are respectively defined as

$$\xi_i^- = \max\{\dot{\theta}_i^-, \mu(\theta_i^- - \theta_i)\}, \qquad \xi_i^+ = \min\{\dot{\theta}_i^+, \mu(\theta_i^+ - \theta_i)\}, \tag{11.6}$$

where $\mu > 0$ is used to scale the feasible region of joint velocity $\dot{\theta}$. As shown in Fig. 11.3, Eq. (11.6) directly uses the actual joint physical limits $\theta^{\pm}$ in the bound constraint of the scheme formulation, which is evidently inapplicable to the presented planar PRJ manipulator. For example, in the presented six-DOF PRJ robot manipulator shown in Fig. 11.1, Joints 2–6 of the robot manipulator each include a safety device (i.e., the photoelectric switch) shown in Fig. 11.4. The LED indicators of the limit positions are "on" (i.e., in red color) under normal circumstances, as shown in the left photo of Fig. 11.4. As further shown in the right photo of Fig. 11.4, if a joint of the manipulator approaches or reaches its physical limit (i.e., the corresponding LED indicator is "off" because the occluder blocks the light beam from

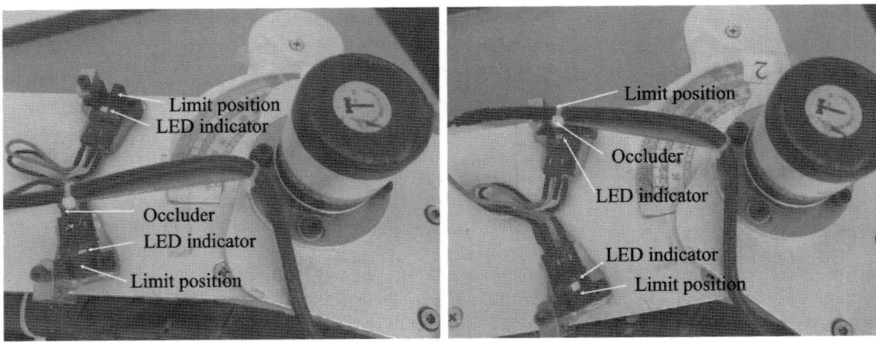

Fig. 11.4 Safety device of the six-DOF PRJ robot manipulator

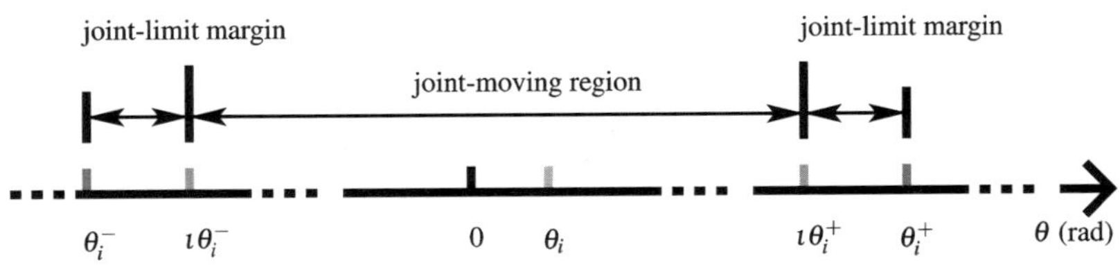

Fig. 11.5 Joint-limit conversion with margins considered for $\theta_i^- < 0 < \theta_i^+$

the transmitter toward the receiver), the joint would be locked by the physical safety device throughout the task duration, resulting in the failure of the end-effector task execution.

Unlike Eq. (11.6), wherein the bound constraint is designed by directly using joint limits, [9] develops the following expressions with some joint-limit margins considered:

$$\xi_i^- = \max\{\dot{\theta}_i^-, \mu(\iota\theta_i^- - \theta_i)\}, \qquad \xi_i^+ = \min\{\dot{\theta}_i^+, \mu(\iota\theta_i^+ - \theta_i)\}, \tag{11.7}$$

where the critical coefficient $\iota \in (0, 1)$ is selected to define a critical region $(\theta_i^-, \iota\theta_i^-]$ or $[\iota\theta_i^+, \theta_i^+)$ as shown in Fig. 11.5, and μ is defined the same as before [e.g., in (11.6) and (2.21)]. This design method can avoid the situation wherein the ith joint is locked near a joint physical limit, where the joint limits satisfy $\theta_i^- < 0 < \theta_i^+$.

However, when the limits of the ith joint satisfy $0 < \theta_i^- < \theta_i^+$, (11.7) may cause a problem wherein the lower bound is expanded [i.e., from θ_i^- to $\iota\theta_i^-$ with $\iota \in (0, 1)$] and is thus invalid. The following expressions are designed to solve this problem:

$$\begin{aligned} \xi_i^- &= \max\{\dot{\theta}_i^-, \mu((2-\iota)\theta_i^- - \theta_i)\}, \\ \xi_i^+ &= \min\{\dot{\theta}_i^+, \mu(\iota\theta_i^+ - \theta_i)\}, \end{aligned} \tag{11.8}$$

of which the geometric interpretation is shown in Fig. 11.6.

In the presented PRJ manipulator, since $\theta_1^- < 0 < \theta_1^+$ and $0 < \theta_i^- < \theta_i^+$ (with $i = 2, 3, 6$), the final joint-limit conversion method for $\xi^- \le \dot{\theta} \le \xi^+$ can be exploited

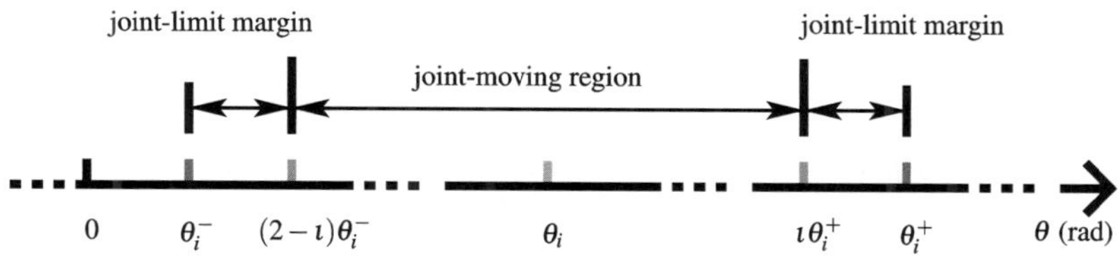

Fig. 11.6 Joint-limit conversion with margins considered for $0 < \theta_i^- < \theta_i^+$

by following the design formulas (11.7) and (11.8):

$$\xi_i^- = \begin{cases} \max\{\dot{\theta}_i^-(\theta_i), \mu(\iota\theta_i^- - \theta_i)\}, & i = 1, \\ \max\{\dot{\theta}_i^-(\theta_i), \mu((2-\iota)\theta_i^- - \theta_i)\}, & i = 2, 3, \ldots, 6, \end{cases}$$
$$\xi_i^+ = \begin{cases} \min\{\dot{\theta}_i^+(\theta_i), \mu(\iota\theta_i^+ - \theta_i)\}, & i = 1, \\ \min\{\dot{\theta}_i^+(\theta_i), \mu(\iota\theta_i^+ - \theta_i)\}, & i = 2, 3, \ldots, 6, \end{cases} \tag{11.9}$$

where μ and ι are defined the same as before, with their values set according to the sensitivity of the safety devices of the robot manipulator and the actual experimental results. In the simulations and experiments, $\iota = 0.88$ and $\mu = 4$. The final bounds $\xi_i^\pm$ [i.e., (11.9)] include $\dot{\theta}_i^\pm(\theta_i)$ and θ_i, so the bounds $\xi_i^\pm$ are time-varying, being functions of joint angles and being in accordance with the PRJ structure.

Therefore, the RMP redundancy-resolution scheme of the robot manipulator can finally be reformulated as the following QP:

$$\text{minimize} \quad \hat{x}^{\mathrm{T}} Q \hat{x}/2 + \hat{q}^{\mathrm{T}} \hat{x}, \tag{11.10}$$
$$\text{subject to} \quad J\hat{x} = \hat{d}, \tag{11.11}$$
$$\xi^- \le \hat{x} \le \xi^+, \tag{11.12}$$
$$\text{with} \quad Q = I, \ \hat{d} = \dot{r} + K\big(r - f(\theta)\big),$$
$$\hat{q} = \lambda\big(\theta(t) - \theta(0)\big),$$

where decision variable vector $\hat{x} = \dot{\theta} \in R^n$. The performance index (11.10) results from the derivation of (11.2). The equality constraint (11.11) expresses a pointwise linear relation between the desired Cartesian velocity $\dot{r}$ and the joint velocity $\dot{\theta}$, with position-error feedback $K(r - f(\theta))$ included. To keep all joint variables within their mechanical ranges, using the bound constraint (11.12) is straightforward and concise. In the present work, the additionally considered subtasks for redundancy resolution are the avoidance of joint physical limits [especially the variable joint-velocity limits depicted in (11.1), (11.5), and (11.12)] and repetitive motion.

Remark 11.1 Nearly all effective optimization schemes for redundancy resolution can be finally reformulated as a standard QP problem subject to one, two, or three kinds of constraints (i.e., equality, inequality, and/or bound constraints). Based on such a standard form, many efficient QP solvers are investigated, including recurrent neural networks (e.g., the ones presented in Chaps. 4 and 5) and numerical algorithms (e.g., the ones presented in Chaps. 6 and 7). Hence, different schemes having the same or similar simple formulations [such as (11.10)–(11.12)] are among the primary advantages of the QP-based redundancy-resolution approach. Based on this point, robotics researchers including the authors of the book have exploited different redundancy-resolution schemes and unified them as a standard QP in the past. Unlike the continuous-time QP solvers (i.e., the recurrent neural networks presented in Chaps. 4 and 5 and applied in Chaps. 8 through 10), in this chapter we exploit the discrete-time QP-solution method that can be easily realized on digital computers, i.e., the 94LVI numerical algorithm presented in Chap. 6.

11.3.3 Discrete-Time QP Solver

In this subsection, the 94LVI numerical computing method (or to say, numerical algorithm) is developed as an efficient QP solver for handling the presented QP problem (11.10)–(11.12) using the following design steps.

The QP problem (11.10)–(11.12) can firstly be converted into a linear variational inequality, i.e., (6.2). Mathematically, we are to find a vector $\hat{y}^* \in \Omega := \{\hat{y} \in R^{n+m} \mid \varsigma^- \leq \hat{y} \leq \varsigma^+\} \subset R^{n+m}$ such that $(\hat{y} - \hat{y}^*)^{\mathrm{T}}(H\hat{y}^* + \hat{p}) \geqslant 0$ $\forall \hat{y} \in \Omega$, where the vector $\hat{y}$ and its upper and lower bounds $\varsigma^{\pm}$ are respectively defined as (with $\varpi \gg 0$ used to replace $+\infty$ numerically):

$$\hat{y} = \begin{bmatrix} \hat{x} \\ u \end{bmatrix}, \qquad \varsigma^+ = \begin{bmatrix} \xi^+ \\ \varpi \end{bmatrix}, \qquad \varsigma^- = \begin{bmatrix} \xi^- \\ -\varpi \end{bmatrix}.$$

In the above definition of $\hat{y}$, its constituting vector $u \in R^m$ is actually a dual decision variable vector defined for the equality constraint (11.11). In addition, the coefficient matrix H and vector $\hat{p}$ are respectively defined as

$$H = \begin{bmatrix} Q & -J^{\mathrm{T}} \\ J & 0 \end{bmatrix}, \qquad \hat{p} = \begin{bmatrix} \hat{q} \\ -\hat{d} \end{bmatrix}.$$

Note that the inequality $(\hat{y} - \hat{y}^*)^{\mathrm{T}}(H\hat{y}^* + \hat{p}) \geqslant 0$ is equivalent to the following piecewise-linear equation (PLE) [10]:

$$\mathscr{P}_\Omega\big(\hat{y} - (H\hat{y} + \hat{p})\big) - \hat{y} = 0, \tag{11.13}$$

where $\mathscr{P}_\Omega(\cdot)$ denotes the projection operator, which is widely used in this book (e.g., in Chaps. 4 through 7). According to [11], such a PLE can be solved by the following 94LVI numerical algorithm. Given any initial vector $\hat{y}^0 \in R^{n+m}$, for iteration index $\kappa = 0, 1, 2, 3, \ldots$, if $\hat{y}^\kappa$ is not a solution of (11.13), the 94LVI numerical iteration formula for $\hat{y}^{\kappa+1}$ is

$$\hat{y}^{\kappa+1} = \hat{y}^\kappa - \hat{\rho}\big(\hat{y}^\kappa\big)\sigma\big(\hat{y}^\kappa\big), \tag{11.14}$$

where $e(\hat{y}^\kappa) = \hat{y}^\kappa - \mathscr{P}_\Omega(\hat{y}^\kappa - (H\hat{y}^\kappa + \hat{p}))$, $\sigma(\hat{y}^\kappa) = (H^{\mathrm{T}} + I)e(\hat{y}^\kappa)$, and $\hat{\rho}(\hat{y}^\kappa) = \|e(\hat{y}^\kappa)\|_2^2/\|\sigma(\hat{y}^\kappa)\|_2^2$. Based on Theorems 1–3 of [11], the sequence $\{\hat{y}^\kappa\}$ generated by the 94LVI numerical algorithm (11.14) globally linearly converges to a solution $\hat{y}^*$, of which the first n elements constitute the optimal solution $\hat{x}^*$ (i.e., the resolved $\dot{\theta}$) for QP (11.10)–(11.12).

11.3.4 Control Process of PRJ Manipulator

The motors of the robot joints are driven by the pulses transmitted from the host computer, so the computed joint variables (i.e., the joint angle θ and joint velocity $\dot{\theta}$) should be converted into pulses per second (PPS), i.e., the actual control input to the joints. For Joint 1 (i.e., the servo motor), the PPS of the servo motor is

$$\mathrm{PPS}_1 = \hat{\tau}\dot{\theta}_1/(2\pi), \tag{11.15}$$

where $\hat{\tau} = 3.2 \times 10^5$ is the parameter related to the hardware system.

For Joints 2 through 6 (i.e., the stepper motors), the structure of the ith joint is shown in Fig. 11.2, and the rotation rate of the ith motor is

$$\tilde{v}_i = \alpha_i\beta_i\dot{\theta}_i\cos\theta_i/\left(\tilde{s}_i\sqrt{\alpha_i^2 + \beta_i^2 + 2\alpha_i\beta_i\sin\theta_i}\right)$$

with $i = 2, 3, \ldots, 6$. As the stepping angles of Joints 2 through 6 are all 0.01π rad and the subdividing multiples are all 32, the PPS for the ith motor and joint ($i = 2, 3, \ldots, 6$) is

$$\mathrm{PPS}_i = \big(2\pi/(0.01\pi/32)\big)\tilde{v}_i = \frac{6400\alpha_i\beta_i\dot{\theta}_i\cos\theta_i}{\tilde{s}_i\sqrt{\alpha_i^2 + \beta_i^2 + 2\alpha_i\beta_i\sin\theta_i}}. \tag{11.16}$$

Generally, the physically constrained RMP scheme proposed in the chapter includes the above-mentioned QP (11.10)–(11.12), the QP solver (11.14), and the control process (11.15)–(11.16).

11.4 Simulative and Experimental Verification

In this section, two different path tracking experiments are demonstrated based on the planar six-DOF PRJ manipulator to substantiate the physical realizability and efficacy of the proposed physically constrained RMP scheme. As a detailed path tracking experiment, the square-path tracking results of nonrepetitive and repetitive motions are both presented for illustration and comparison. As a supplementary experiment, the tracking of the B-shaped path and circular path are shown to further validate the effectiveness on more complex tracking tasks. The position-error analysis illustrates the accuracy of the proposed scheme. Note that the robot is of six-DOF and only the end-effector position is considered in the present work, so the dimension of Jacobian matrix is 2×6, and the degrees of redundancy are four.

Table 11.2 Joint displacement when the drift-free criterion is not considered (i.e., $\lambda = 0$)

#	$\theta(0)$		$\theta(40)$		$\theta(40) - \theta(0)$	
	(rad)	(degree)	(rad)	(degree)	(rad)	(degree)
θ_1	+0.5236	+30.0000	+0.5002	+28.6613	−0.0234	−1.3407
θ_2	+0.2618	+15.0000	+0.2960	+16.9591	+0.0342	+1.9595
θ_3	+0.2618	+15.0000	+0.2920	+16.7317	+0.0302	+1.7303
θ_4	+0.2618	+15.0000	+0.2114	+12.1122	−0.0504	−2.8877
θ_5	+0.2618	+15.0000	+0.2474	+14.1756	−0.0144	−0.8251
θ_6	+0.2618	+15.0000	+0.2861	+16.3895	+0.0243	+1.3923

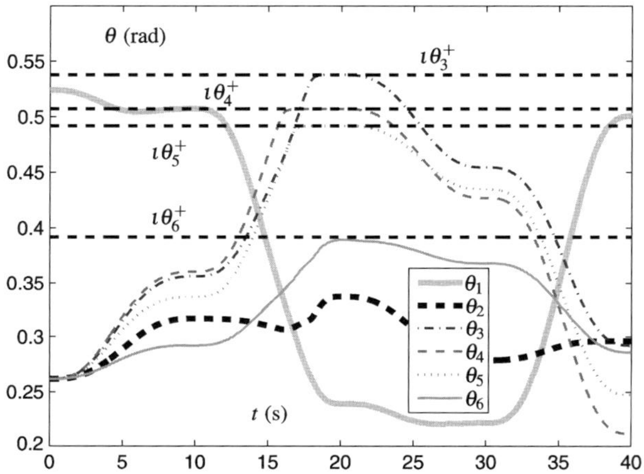

Fig. 11.7 Joint-angle profiles when the planar six-DOF PRJ manipulator's end-effector tracks a square path without considering the drift-free criterion (i.e., $\lambda = 0$)

11.4.1 Square-Path Tracking

In this subsection, the end-effector of the planar six-DOF PRJ robot manipulator is expected to track a square path with the side length being 18 cm (i.e., 0.18 m) for the two square-path tracking experiments. The initial joint state $\theta(0) = [\pi/6, \pi/12, \pi/12, \pi/12, \pi/12, \pi/12]^{\mathrm{T}}$ rad (i.e., $[30, 15, 15, 15, 15, 15]^{\mathrm{T}}$ in degrees), is shown in Table 11.2. In addition, the task duration T_{d} is 40 s.

11.4.1.1 Nonrepetitive Motion

When the drift-free coefficient $\lambda = 0$ is set, i.e., without considering the drift-free criterion, scheme (11.2)–(11.5) reduces to the minimum velocity norm scheme, which may not guarantee the repetitive motion of the redundant robot. Figure 11.7 shows the profiles of the joint angles when the end-effector tracks the square path without considering the drift-free criterion (i.e., $\lambda = 0$). From Fig. 11.7 we can see that the final state of joint-angle vector $\theta(40)$ of the manipulator does not coincide well with its initial state of joint-angle vector $\theta(0)$, i.e., $\theta(40) \neq \theta(0)$, which

Table 11.3 Joint displacement when the drift-free criterion is activated (i.e., $\lambda = 4$)

#	$\theta(0)$		$\theta(40)$		$\theta(40) - \theta(0)$
	(rad)	(degree)	(rad)	(degree)	(rad)
θ_1	+0.5236	+30.0000	+0.52360004	+30.00007241	$+1.2638 \times 10^{-6}$
θ_2	+0.2618	+15.0000	+0.26179927	+14.99999352	-1.1307×10^{-7}
θ_3	+0.2618	+15.0000	+0.26179864	+14.99995694	-7.5149×10^{-7}
θ_4	+0.2618	+15.0000	+0.26179848	+14.99994817	-9.0456×10^{-7}
θ_5	+0.2618	+15.0000	+0.26179859	+14.99995453	-7.9353×10^{-7}
θ_6	+0.2618	+15.0000	+0.26179899	+14.99997727	-3.9666×10^{-7}

is also shown in Table 11.2 (particularly, the last column). In engineering applications, this situation may induce a problem wherein the behavior of the manipulator is hard to predict. Besides, readjusting the configuration of the manipulator through self-motion is less efficient [6, 7].

11.4.1.2 Repetitive Motion

To finish the closed-path tracking task on the PRJ robot manipulator in a repetitive manner, the physically constrained RMP scheme subject to VJVL [i.e., QP (11.10)–(11.12)] and the corresponding 94LVI numerical algorithm [i.e., (11.14)] are employed to solve the nonrepetitive motion problem. That is, the joint drift criterion is activated with $\lambda = 4$ in the simulation and experiment of this subsection. The corresponding simulative and experimental results are illustrated in Figs. 11.8, 11.9 and 11.10, Table 11.3, and Figs. 11.11 and 11.12.

Specifically, it is worth mentioning that the motors of the robot manipulator are driven by the pulses transmitted from the host computer, i.e., the actual control input to the joints in the experiment. The values of joint variables (i.e., θ and $\dot{\theta}$) can be calculated by exploiting the 94LVI numerical algorithm (11.14) that solves QP (11.10)–(11.12). The PPS are then generated according to (11.15) and (11.16). Evidently, such calculation results expressed in floating-point format have to be rounded, i.e., from floating-point numbers to integers. The latter are used as the actual PPS (shown in Fig. 11.8) to control the manipulator.

The actual task execution of the manipulator hardware system synthesized by the physically constrained RMP scheme subject to VJVL is then shown in Fig. 11.9, which demonstrates that the task is performed well. The computer-simulation results, shown correspondingly in Fig. 11.10, confirm the findings evidently. The left graph of Fig. 11.10 shows the simulated joint-motion trajectories and their proportions of the manipulator. From the graph we see that the final state coincides well with the initial one. Besides, as substantiated by the right graph of Fig. 11.10, after completing the end-effector tracking task, the final joint state $\theta(40)$ returns to the initial joint state $\theta(0)$. The detailed numerical data are further shown in Table 11.3. Comparatively speaking, the joint displacement $\theta(40) - \theta(0)$ shown in Table 11.3

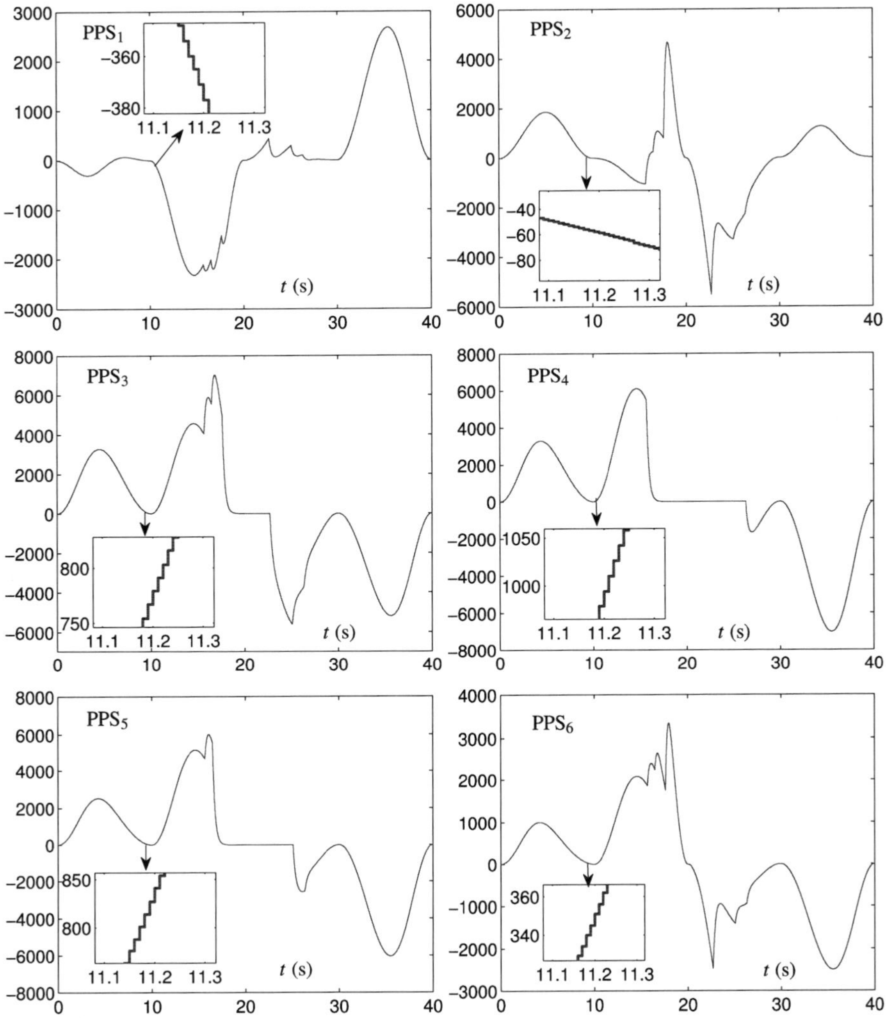

Fig. 11.8 PPS transmitted to the motors of the joints when the end-effector tracks the square path with the drift-free criterion activated (i.e., $\lambda = 4$)

is significantly smaller than that in the nonrepetitive motion situation (shown in Table 11.2) because the drift-free criterion is considered and activated. In other words, the joint displacement $\theta(40) - \theta(0)$ is tiny and acceptable in practice. In addition, the joint angles are all kept in their limited ranges during the task execution, as shown in the right graph of Fig. 11.10.

Furthermore, Fig. 11.11 presents the ith joint velocity $\dot{\theta}_i$, its corresponding limits $\dot{\theta}_i^{\pm}$, and the designed bounds $\xi_i^{\pm}$, with $i = 1, 2, \ldots, 6$, verifying that the joint-velocity limits $\dot{\theta}_i^{\pm}$ and bounds $\xi_i^{\pm}$ are time-varying and change with the end-effector and joints movement, i.e., the variable joint-velocity limits. These figures substanti-

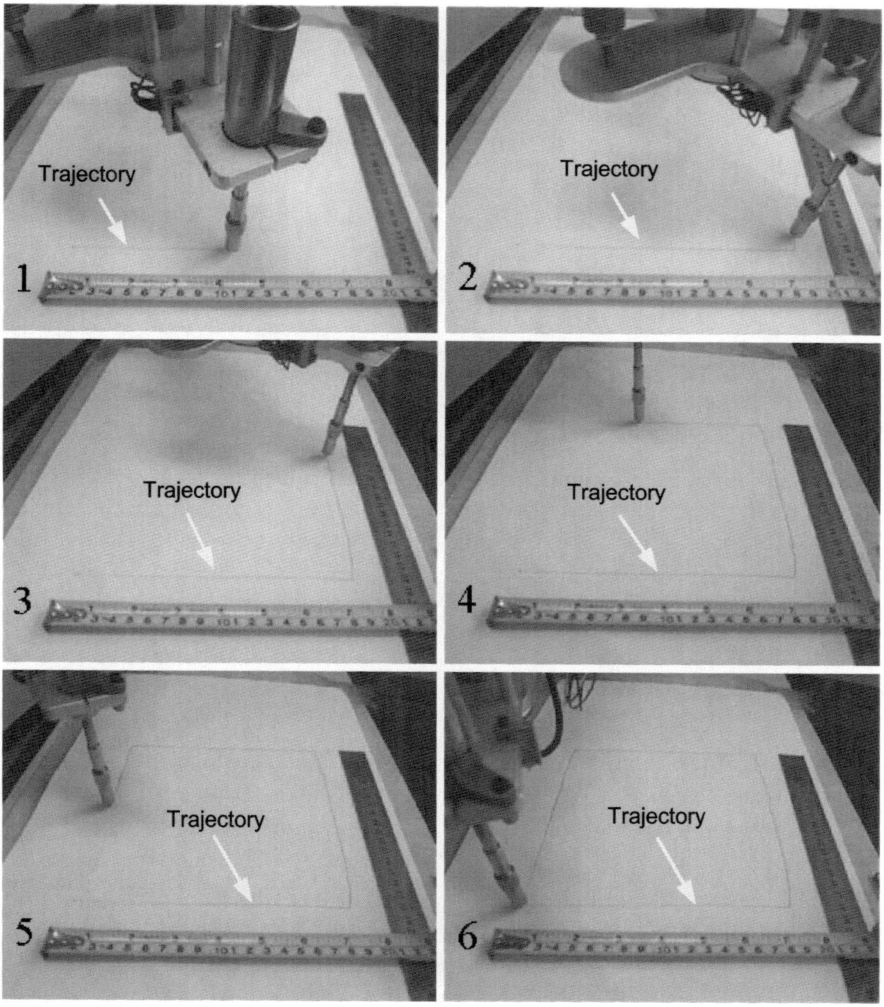

Fig. 11.9 Snapshots of the actual task execution of the planar six-DOF PRJ robot manipulator synthesized by the proposed physically constrained RMP scheme subject to VJVL and with the drift-free criterion activated (i.e., $\lambda = 4$) when the end-effector tracks the square path

ate that the proposed physically constrained RMP scheme subject to VJVL is practically feasible and effective. Moreover, Fig. 11.12 presents the results of the tracking error analysis. In detail, the desired path and the tracking trajectory of the end-effector are demonstrated in the left graph of Fig. 11.12. The end-effector trajectory is shown to coincide well with the desired path. In the right graph of Fig. 11.12, the end-effector position error $\tilde{e}$ is correspondingly shown, illustrating the tiny deviations of the end-effector trajectory $f(\theta)$ from the desired path r in the X and Y axes of the base frame (i.e., $\tilde{e}_{\mathrm{X}}$ and $\tilde{e}_{\mathrm{Y}}$, respectively). As shown in the graph, the maximal position error is less than 8.0×10^{-5} m, illustrating that the end-effector trajectory

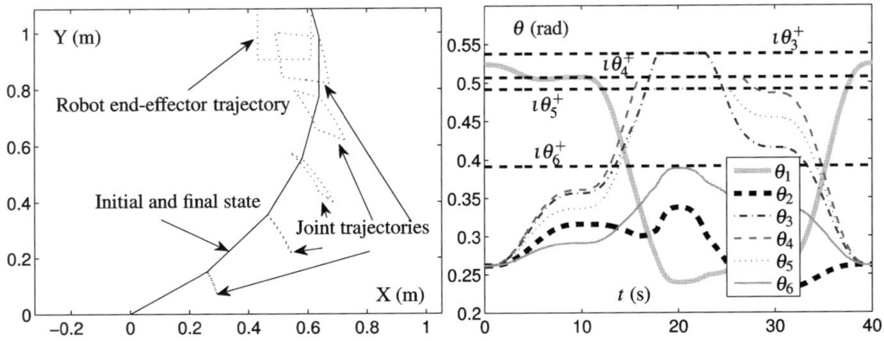

Fig. 11.10 The end-effector of the planar six-DOF PRJ manipulator tracks the square path with the drift-free criterion activated (i.e., $\lambda = 4$)

of the six-DOF PRJ robot manipulator synthesized by the proposed physically constrained RMP scheme in the two-dimensional work-plane is sufficiently close to the desired square path. Hence, the square-path tracking task is completed successfully and accurately.

11.4.2 B-Shaped Path Tracking

For demonstrating the effectiveness of tracking a more complex path, the end-effector of the planar six-DOF PRJ robot manipulator is expected to trace a B-shaped path (which has both curve and straight-line segments). The side length of the B-shaped path is 6 cm, and the task duration is 15 s. The snapshots of the actual task execution when the robot tracks the B-shaped path are shown in the left photo of Fig. 11.13. The right photo of Fig. 11.13 shows the measurement of the experimental result of the B-shaped path tracking, which demonstrates that the task is fulfilled very well. The tiny position error (i.e., less than 1.5×10^{-5} m) for the tracking error analysis can be observed from Fig. 11.14. Both of the actual experiment and simulation substantiate that the proposed physically constrained RMP scheme is physically realizable, effective, and accurate in a relatively wide range of practical applications.

11.4.3 Circular Path Tracking

To further demonstrate the physical realizability and effectiveness of the proposed redundancy-resolution method (i.e., the RMP scheme subject to VJVL and the 94LVI numerical algorithm), a circular-path tracking experiment is further performed and presented. The diameter of the circle is 2.5 cm, and the task duration T_d is 10 s. For comparison and demonstration, $\lambda = 4$ is used, with the initial state

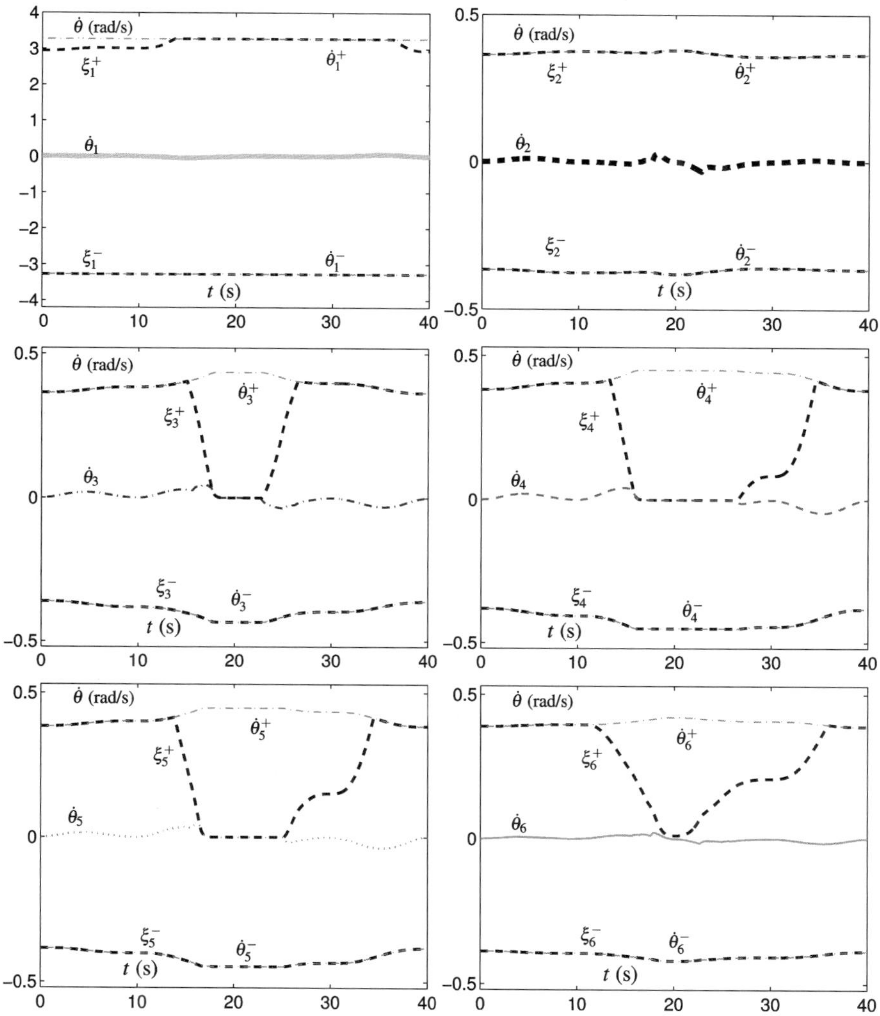

Fig. 11.11 Joint velocity $\dot{\theta}$, its limits $\dot{\theta}^{\pm}$, and the designed bounds $\xi^{\pm}$ of the planar six-DOF PRJ robot manipulator synthesized by the proposed physically constrained RMP scheme with drift-free coefficient $\lambda = 4$ when the end-effector tracks a square path, where joint velocity $\dot{\theta}$ is within its limits $\dot{\theta}^{\pm}$ and bounds $\xi^{\pm}$

$\theta(0)$ and other design parameters (e.g., K, μ, and ι) being set the same as before. The snapshots of the actual task execution are illustrated in Fig. 11.15. The figure indicates that the circle is completed successfully, substantiating the physical realizability and effectiveness of the proposed RMP scheme.

In summary, the preceding experiments have substantiated the physical realizability and effectiveness of the proposed physically constrained RMP scheme subject to VJVL for solving the nonrepetitive problem of redundant robot manipulators. The position error analysis has further validated the accuracy of this scheme.

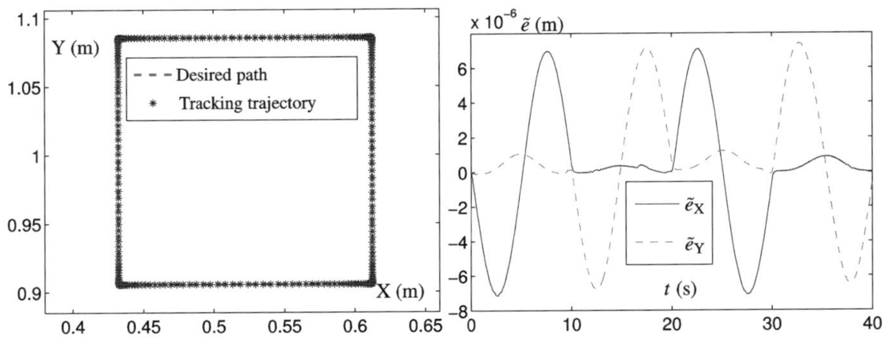

Fig. 11.12 Desired path, tracking trajectory, and position error of the end-effector of the six-DOF PRJ manipulator synthesized by the proposed physically constrained RMP scheme with drift-free coefficient $\lambda = 4$ when the end-effector tracks the square path

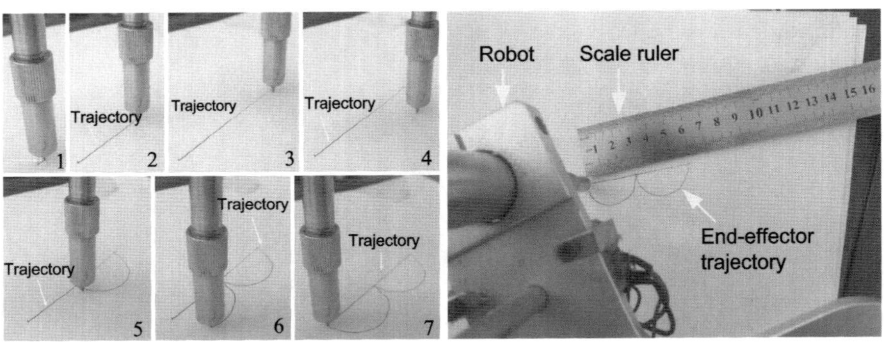

Fig. 11.13 Snapshots of the actual task execution and measurement of the experimental result of the planar six-DOF PRJ robot manipulator synthesized by the proposed physically constrained RMP scheme with drift-free coefficient $\lambda = 4$ when the end-effector tracks a B-shaped path

11.5 Chapter Conclusions

As illustrated through experiments based on a six-DOF PRJ redundant robot manipulator, the joint-drift phenomenon has been remedied effectively (i.e., the repetitive motion planning and control have been achieved successfully). The novel physically constrained RMP scheme subject to VJVL (derived with PRJ) has been proposed and solved by using the PLE-based numerical method (i.e., the 94LVI numerical algorithm). More importantly, this RMP scheme has been simulated and implemented on a practical six-DOF PRJ robot manipulator. The results of comparative simulations and hardware experiments performed on the robot manipulator with square-path, B-shaped-path, and circular-path tracking tasks have further demonstrated the physical realizability and effectiveness of this physically constrained RMP scheme and the 94LVI numerical algorithm. Moreover, the end-effector position-error analysis has validated the accuracy of the physically constrained RMP scheme and the 94LVI numerical algorithm.

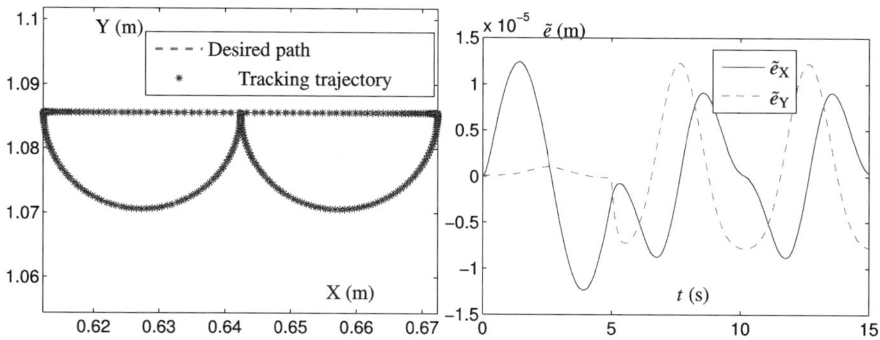

Fig. 11.14 Desired path, tracking trajectory, and position error of the end-effector of the planar six-DOF robot manipulator synthesized by the proposed physically constrained RMP scheme with drift-free coefficient $\lambda = 4$ when the end-effector tracks a B-shaped path

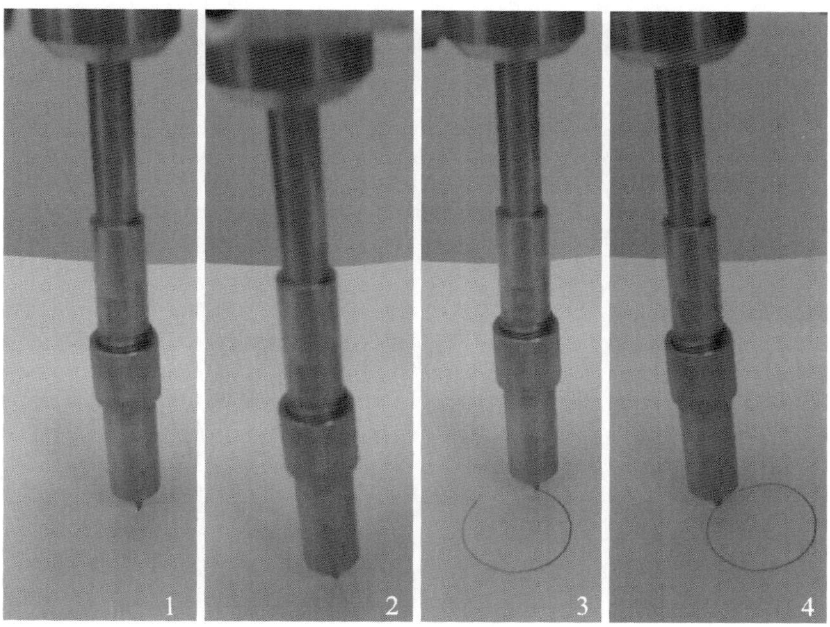

Fig. 11.15 Snapshots of the actual task execution of the planar six-DOF PRJ robot manipulator synthesized by the proposed physically constrained RMP scheme subject to VJVL when the end-effector tracks a circular path

At the end of the chapter, the part, and the book, it is worth pointing out that the E47 numerical algorithm and the recurrent neural networks presented in this book (specifically in Chaps. 4, 5, and 7) are also effective and efficient on the robotic RMP problem solving and thus applicable to the hardware experiment. However, in view of the results' similarity, content conciseness, and space limitation, the corresponding experimental results of using the E47, DNN, LVI-PDNN, and S-LVI-

PDNN have been omitted. As a future research direction, more experiments and applications of the RMP scheme and QP solvers are left to readers to develop and investigate on different physical robot manipulators.

References

1. Agrawal SK, Veeraklaew T (1998) Designing robots for optimal performance during repetitive motion. IEEE Trans Robot Autom 14(5):771–777
2. Liu TS, Lee WS (1999) A repetitive learning method based on sliding mode for robot control. J Dyn Syst Meas Control 122(1):40–48
3. Shamir T, Yomdin Y (1988) Repeatability of redundant manipulators: mathematical solution of the problem. IEEE Trans Autom Control 33(11):1004–1009
4. Luca AD, Mattone R, Oriolo G (1997) Control of redundant robots under end-effector commands: a case study in underactuated systems. Int J Appl Math Comput Sci 7(2):101–127
5. Bowling A, Harmeyer S (2010) Repeatable redundant manipulator control using nullspace quasivelocities. J Dyn Syst Meas Control 132(3):1–11
6. Zhang Y, Lv X, Li Z, Yang Z, Chen K (2008) Repetitive motion planning of PA10 robot arm subject to joint physical limits and using LVI-based primal–dual neural network. Mechatronics 18(9):475–485
7. Zhang Y, Tan Z, Chen K, Yang Z, Lv X (2009) Repetitive motion of redundant robots planned by three kinds of recurrent neural networks and illustrated with a four-link planar manipulator's straight-line example. Robot Auton Syst 57(6–7):645–651
8. Zhang Y, Ma W, Cai B (2009) From Zhang neural network to Newton iteration for matrix inversion. IEEE Trans Circuits Syst I, Regul Pap 56(7):1405–1415
9. Zhang Y, Wang J, Xia Y (2003) A dual neural network for redundancy resolution of kinematically redundant manipulators subject to joint limits and joint velocity limits. IEEE Trans Neural Netw 14(3):658–667
10. Kinderlehrer D, Stampcchia G (1980) An introduction to variational inequalities and their applications. Academic Press, New York
11. He B (1994) A new method for a class of linear variational inequalities. Math Program 66(1–3):137–144

Appendix A
Proof of Extension-I Performance Index

In this appendix, we employ the well-known gradient-descent approach to analyze and prove the repetitive-motion performance index exploited in the extension-I optimization scheme.

First, to achieve the RMP purpose of redundant robot manipulators, we can define the following scalar-valued norm-based error function (or to say, a variant of conventional energy function):

$$\hat{\varepsilon} = \ln\big(\|\theta - \theta(0)\|_2\big) = \frac{1}{2}\ln\big(\|\theta - \theta(0)\|_2^2\big).$$

Evidently, the function $\ln(x)$ is an increasing function. Thus, the minimum of such an error function $\hat{\varepsilon}$ is obtained when $\|\theta - \theta(0)\|_2^2$ reaches its minimum point [i.e., $\theta = \theta(0)$, which implies that the RMP purpose is achieved]. However, it is worth pointing out that, in engineering applications, there always exists intrinsic error between the final state and initial state, e.g., 10^{-5}.

Secondly, by using the gradient-descent approach, we can set

$$\dot{\theta} = -\lambda\frac{\partial\hat{\varepsilon}}{\partial\theta} = \frac{-\lambda(\theta - \theta(0))}{\|\theta - \theta(0)\|_2^2}, \tag{A.1}$$

where the design parameter $\lambda > 0$. Such a design idea is to make the error function $\hat{\varepsilon}$ evolve along its negative descent direction until the minimum point is reached [i.e., to make $\theta(t)$ approach to $\theta(0)$] and then to achieve the RMP purpose. Since $\|\theta - \theta(0)\|_2^2$ is a scalar, (A.1) can be reformulated as

$$\|\theta - \theta(0)\|_2\dot{\theta} = \frac{-\lambda(\theta - \theta(0))}{\|\theta - \theta(0)\|_2},$$

which can be rewritten as

$$\|\theta - \theta(0)\|_2\dot{\theta} + \frac{\lambda(\theta - \theta(0))}{\|\theta - \theta(0)\|_2} = 0. \tag{A.2}$$

Y. Zhang, Z. Zhang, *Repetitive Motion Planning and Control of Redundant Robot Manipulators*, DOI 10.1007/978-3-642-37518-7,

Inasmuch as the other factors like the end-effector motion requirement and joint physical limits have to be considered in the scheme formulation, the dynamic equation (A.2) can only be achieved theoretically. Thus, minimizing the following performance index appears to be more feasible for the RMP of redundant manipulators, rather than forcing (A.2) directly in its exact form:

$$\frac{1}{2}\left\| \|\theta - \theta(0)\|_2 \dot{\theta} + \frac{\lambda(\theta - \theta(0))}{\|\theta - \theta(0)\|_2} \right\|_2^2. \tag{A.3}$$

By defining $\hat{c} := \lambda(\theta - \theta(0))$ and expanding (A.3) we can yield

$$\frac{1}{2}\left(\|\theta - \theta(0)\|_2 \dot{\theta} + \frac{\hat{c}}{\|\theta - \theta(0)\|_2} \right)^{\mathrm{T}} \left(\|\theta - \theta(0)\|_2 \dot{\theta} + \frac{\hat{c}}{\|\theta - \theta(0)\|_2} \right). \tag{A.4}$$

Since the extension-I optimization scheme is resolved at the joint-velocity level and the decision variable vector is joint velocity $\dot{\theta}$, similar to the analysis given in Sect. 2.2, we can have the following performance index derived from (A.4):

$$\frac{1}{2}\dot{\theta}^{\mathrm{T}} W \dot{\theta} + \hat{c}^{\mathrm{T}} \dot{\theta}, \tag{A.5}$$

where $W := \|\theta - \theta(0)\|_2^2 I$. Evidently, (A.5) is exactly the performance index exploited in the extension-I optimization scheme. The proof is thus completed. □

Appendix B
Simulations of Extension-I Scheme

Differing from the most preferred quadratic-form RMP scheme (2.2)–(2.5), the extension-I optimization scheme (2.7)–(2.10) is proposed in Sect. 2.3.1 and simulatively investigated in this appendix for online RMP of redundant robot manipulators. Such an RMP scheme (2.7)–(2.10) also takes into account the avoidance of joint physical limits (e.g., joint-angle limits and joint-velocity limits) and aims at remedying the so-called joint-angle drift problem. Note that, similar to the most preferred quadratic-form RMP scheme (2.2)–(2.5) and other optimization schemes, the extension-I optimization scheme (2.7)–(2.10) can also be reformulated and unified as a QP problem, which is shown in Table 2.1.

B.1 PUMA560 Robot Manipulator

The PUMA560 robot is a spatial manipulator having six joints [1–3]. When we consider only the positioning of the end-effector, the PUMA560 robot becomes a functionally redundant manipulator and has three redundant DOF.

B.1.1 Circular Path Tracking

In this section, the manipulator's end-effector is expected to track a circle with the radius being 0.1 m. Without loss of generality, we set task duration $T = 10$ s, design parameter $\lambda = 4$, and initial state $\theta = [0, 0, 0, 0, 0, 0]^{\mathrm{T}}$ rad. Figure B.1 shows the computer-simulation results using the extension-I optimization scheme (2.7)–(2.10) [also termed QP-based RMP scheme (2.7)–(2.10) or quadratic-form RMP scheme (2.7)–(2.10)].

Specifically, Fig. B.1(a) shows the 3-D motion trajectories of the PUMA560 robot manipulator. From it we see that the final joint state (or to say, robot configuration) is consistent with its initial one, which achieves the desired RMP purpose.

Y. Zhang, Z. Zhang, *Repetitive Motion Planning and Control of Redundant Robot Manipulators*, DOI 10.1007/978-3-642-37518-7,

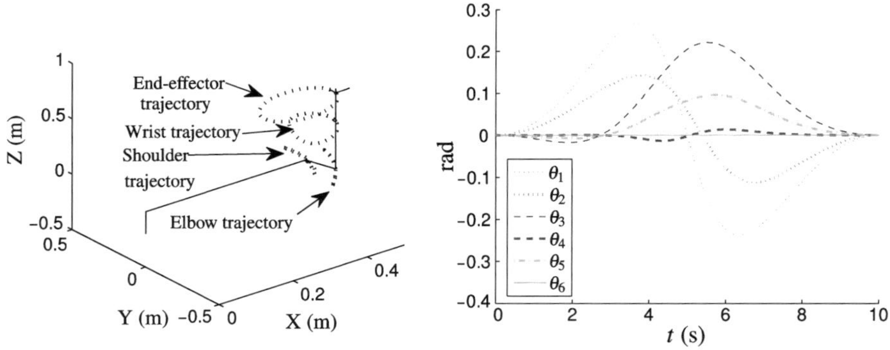

(a) Motion trajectories (b) Joint-angle profiles

Fig. B.1 Motion trajectories and joint-angle profiles of PUMA560 robot manipulator tracking the given circular path

It is worth pointing out that the maximal position error of the end-effector is less than 4×10^{-7} m. Figure B.1(b) further shows the transient trajectories of joint angle. Evidently, the joint variables return to their initial values. Thus, the proposed extension-I optimization scheme is repetitive for the spatially working PUMA560 robot manipulator.

B.1.2 Straight-Line Path Tracking

For the purposes of further demonstrating the proposed RMP scheme, the PUMA560 manipulator's end-effector is expected to move forward and then backward along a straight-line segment with the length being 1.125 m. Without loss of generality, we set task duration $T = 20$ s, design parameter $\lambda = 4$, and initial configuration $\theta = [0, 0, 0, 0, 0, 0]^{\mathrm{T}}$ rad. Computer-simulation results are shown in Fig. B.2.

As seen from Fig. B.2(a), after the end-effector completing the straight-line path tracking task, the final joint state returns to its initial one, which achieves the expected RMP purpose. In addition, the maximal position error of the end-effector is less than 3×10^{-6} m. Figure B.2(b) further shows the transient trajectories of joint angle, from which we know that the joint variables return to their initial states. Thus, the nonrepetitive problem is solved by the proposed RMP scheme as well.

In summary, the above computer-simulation results based on the PUMA560 robot manipulators have demonstrated the efficacy and superiority of the proposed QP-based RMP scheme.

B.2 Four-Link Redundant Planar Manipulator

The four-link redundant planar manipulator is of four DOF [4] and has two redundant degrees with respect to the two-dimensional operational plane (because

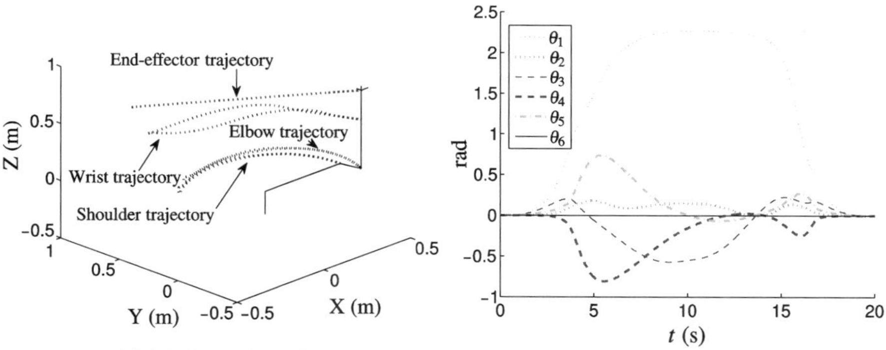

(a) Motion trajectories (b) Joint-angle profiles

Fig. B.2 Motion trajectories and joint-angle profiles of PUMA560 robot manipulator tracking the given straight-line path

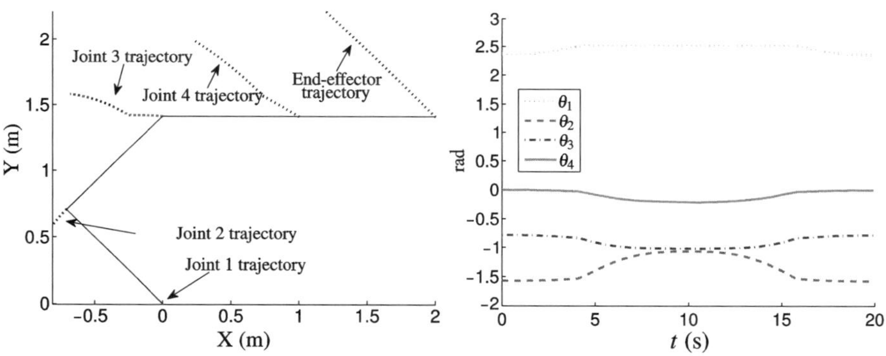

(a) Motion trajectories (b) Joint-angle profiles

Fig. B.3 Motion trajectories and joint-angle profiles of four-link planar redundant robot manipulator tracking the given straight-line path

we only consider the end-effector position). The end-effector is expected to move forward and then backward along a straight-line segment with initial state $\theta(0) = [3\pi/4, -\pi/2, -\pi/4, 0]^{\mathrm{T}}$ rad. Computer-simulation results are shown in Fig. B.3.

As seen from Fig. B.3(a), after the end-effector completing the straight-line path tracking task, the final joint state returns to its initial one. In addition, the maximal position error of the end-effector is less than 1.4×10^{-5} m. As seen from Fig. B.3(b), the final values of joint variables of the four-link redundant planar manipulator are consistent with their corresponding initial values. Thus, the proposed QP-based RMP scheme is effective on solving the nonrepetitive problem.

In summary, the above computer-simulation results based on the PUMA560 and the four-link planar manipulator all have demonstrated the efficacy and superiority of the proposed QP-based RMP scheme, i.e., the extension-I optimization scheme (2.7)–(2.10).

References

1. Zhang Y (2002) Analysis and design of recurrent neural networks and their applications to control and robotic systems. PhD dissertation, Chinese University of Hong Kong, Hong Kong
2. Zhang Y, Ge SS, Lee TH (2004) A unified quadratic programming based dynamical system approach to joint torque optimization of physically constrained redundant manipulators. IEEE Trans Syst Man Cybern, Part B, Cybern 34(5):2126–2132
3. Zhang Y, Wang J (2002) A dual neural network for constrained torque optimization of kinematically redundant manipulators. IEEE Trans Syst Man Cybern, Part B, Cybern 32(5):654–662
4. Zhang Y, Wang J, Xia Y (2003) A dual neural network for redundancy resolution of kinematically redundant manipulators subject to joint limits and joint velocity limits. IEEE Trans Neural Netw 14(3):658–667

Appendix C
Simulations of Extension-II Scheme

In this appendix, focusing on the extension-II optimization scheme (2.12)–(2.15), multiple types of performance indices (termed the most preferred RMP index and its simple extensions) are investigated comparatively for online RMP of redundant manipulators [1]. In addition, physical constraints such as joint limits and joint-velocity limits are taken into consideration in these related scheme formulations, and these schemes can finally be rewritten as QPs (as shown in Table 2.1). These schemes are then applied and simulated based on three different types of robot arms, which shows that the most preferred RMP scheme is effective, whereas its simple extensions (including the extension-II optimization scheme) may not work applicably.

C.1 PA10 Robot Manipulator

In this section, computer simulations are performed based on the 7-DOF PA10 robot manipulator again, of which the parameters and limits are mentioned in Chap. 1. Two simulation examples are shown as follows (i.e., a circular path tracking example and a straight-line path tracking example).

C.1.1 Circular Path Tracking

In this circular example, the motion trajectory of the PA10 end-effector is expected to be a circle with radius 0.2 m and the revolute angle about X-axis being $\pi/6$ rad. The task duration is 10 s, and initial joint state $\theta(0) = [0, \pi/4, 0, 2\pi/3, 0, -\pi/4, 0]^{\mathrm{T}}$ in radians.

First, let us consider the situation of no drift-free criterion; i.e., the performance index is $\dot{\theta}^{\mathrm{T}}\dot{\theta}/2$ by forcing $\lambda = 0$ in (2.2). As seen from the upper graph of Fig. C.1 with $\mu = 1$, the solution is not repetitive in the sense that the final state of the PA10

Y. Zhang, Z. Zhang, *Repetitive Motion Planning and Control of Redundant Robot Manipulators*, DOI 10.1007/978-3-642-37518-7,

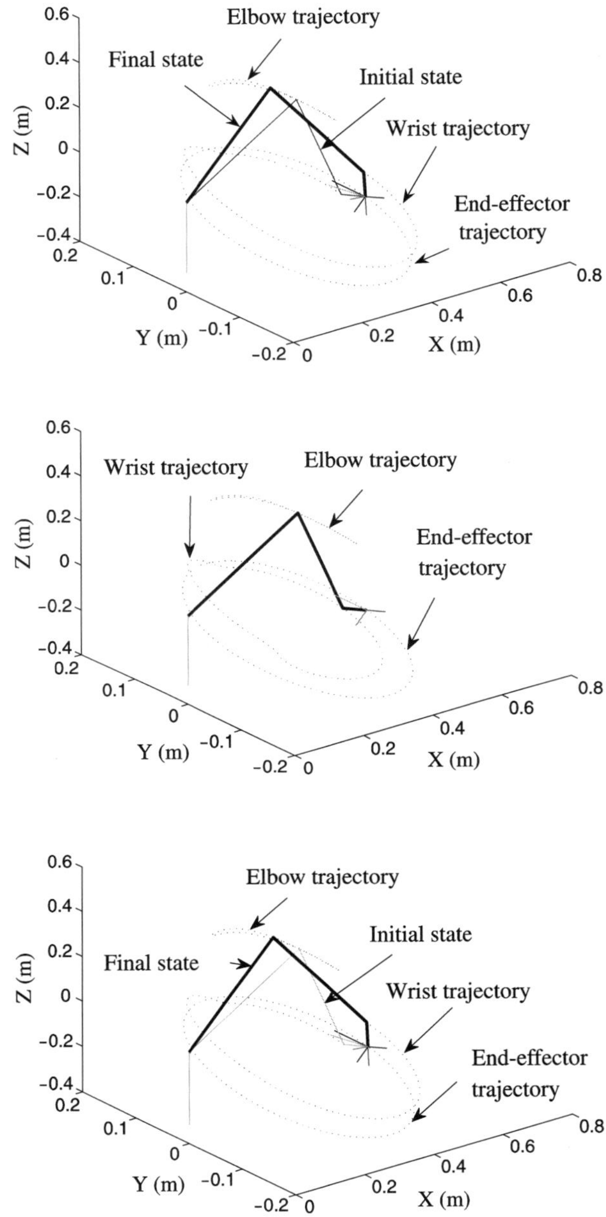

Fig. C.1 Performances of different indices [i.e., minimum-velocity performance index $\dot{\theta}^{\mathrm{T}}\dot{\theta}/2$, the most preferred RMP performance index (2.2), and extension-II performance index (2.12)] on PA10 robot manipulator with its end-effector tracking a circular path, respectively

manipulator does not coincide with its initial state. Therefore, such a solution is thought to have a joint-angle drift problem (or to say, nonrepetitive problem).

Second, let us consider the inclusion of the most preferred RMP performance index (2.2) with $\lambda = 4$ and $\mu = 1$. The middle graph of Fig. C.1 shows the corresponding simulated result, which demonstrates that the solution is repetitive [as the final and initial PA10 states coincide well with each other].

Third, let us consider the extension-II performance index (2.12) and its corresponding scheme (2.12)–(2.15) with $\mu = 1$. As seen from the lower graph of Fig. C.1, there evidently still exists a joint-angle drift problem (i.e., nonrepetitive problem). In other words, in this PA10 circular example, extension-II performance index (2.12) cannot achieve the RMP purpose, though it appears to be capable.

C.1.2 Straight-Line Path Tracking

In this second example of PA10, the motion trajectory of the end-effector is to move forward and then backward along a straight-line segment. The task duration is 7 s, the line-segment length is 2.5 m, and the initial state $\theta(0) = [\pi/4, \pi/4, \pi/4, \pi/4, \pi/4, \pi/4, 0]^{\mathrm{T}}$ rad. The angles of the desired straight-line segment made with XY, YZ, and ZX planes are $\pi/4$ rad, $\pi/6$ rad, and $\pi/6$ rad, respectively.

First, let us consider the situation of no drift-free criterion and $\mu = 4$. The result is shown in the upper graph of Fig. C.2, and evidently the PA10's final state does not coincide with its initial state. That is, a so-called joint-angle drift problem occurs.

Second, let us consider the inclusion of the most preferred RMP performance index (2.2) with $\lambda = 4$ and $\mu = 4$. The middle graph of Fig. C.2 shows the corresponding result, i.e., the solution is now repetitive in the sense that the final PA10 state equals the initial state.

Third, let us consider the extension-II performance index (2.12) and its corresponding scheme (2.12)–(2.15) with $\mu = 4$ in this PA10 straight-line example. The result is shown in the lower graph of Fig. C.2. Evidently, such an extended performance index could not achieve the RMP purpose either, and its performance is similar to that of $\dot{\theta}^{\mathrm{T}}\dot{\theta}/2$.

C.2 PUMA560 Robot Manipulator

In this section, computer simulations are further conducted based on the PUMA560 robot manipulator [3, 4, 6] to verify the theoretical analysis presented in Sect. 2.3.2 and the PA10 observations [e.g., extension-II performance index (2.12) may not work practically]. In addition, it is worth noting that the QP solution in all simulations of this appendix is actually solved via the LVI-based primal–dual neural network (which is presented in Sect. 5.2.5 and [1, 2, 4]).

C.2.1 Circular Path Tracking

In this example, the end-effector of the PUMA560 manipulator is to move along a circle in the three-dimensional workspace. The upper graph of Fig. C.3 shows the

Fig. C.2 Performances of different indices [i.e., minimum-velocity performance index $\dot{\theta}^{\mathrm{T}}\dot{\theta}/2$, the most preferred RMP performance index (2.2), and extension-II performance index (2.12)] on PA10 robot manipulator with its end-effector tracking a straight-line path, respectively

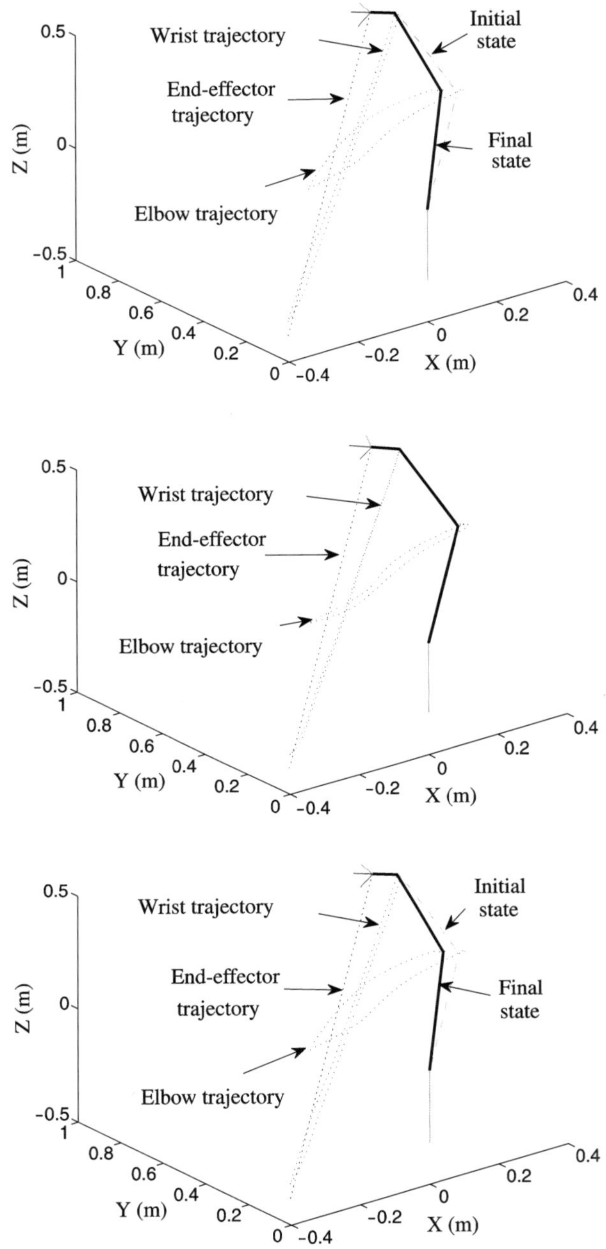

result by using usual minimum-velocity performance index $\dot{\theta}^{\mathrm{T}}\dot{\theta}/2$, and the middle graph of Fig. C.3 shows the result by using the most preferred RMP performance index (2.2). Evidently, by exploiting the most preferred RMP performance index (2.2), the repetitive motion planning can definitely be achieved for the PUMA560 circular

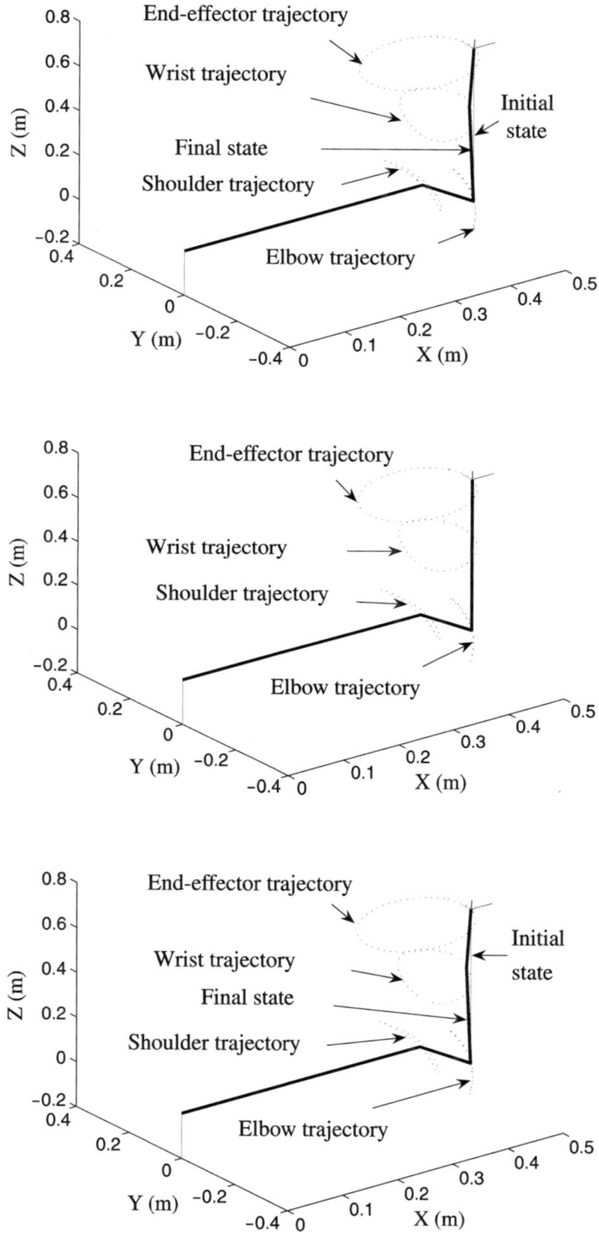

Fig. C.3 Performances of different indices [i.e., minimum-velocity performance index $\dot{\theta}^{\mathrm{T}}\dot{\theta}/2$, the most preferred RMP performance index (2.2), and extension-II performance index (2.12)] on PUMA560 robot manipulator with its end-effector tracking a circular path, respectively

example, whereas the minimum-velocity performance index may not achieve the RMP purpose.

For comparison and illustration, the extension-II performance index (2.12) is also exploited with joint physical limits considered. However, as the lower graph of Fig. C.3 shows, such a simulated solution is not cyclic (or to say, not repetitive).

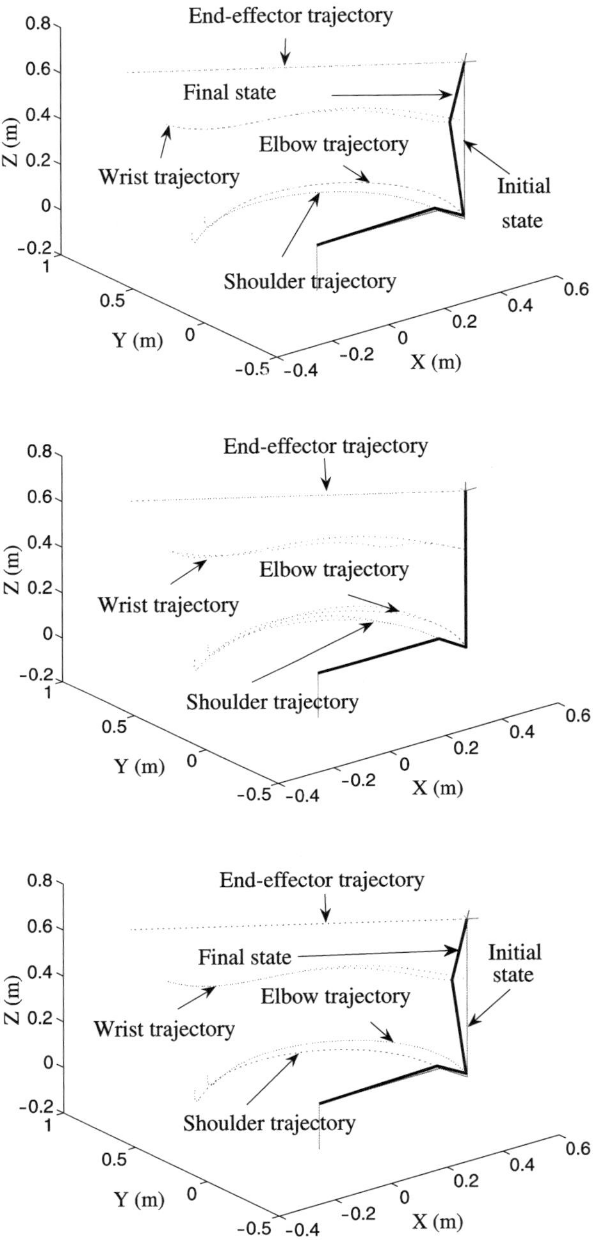

Fig. C.4 Performances of different indices [i.e., minimum-velocity performance index $\dot{\theta}^{\mathrm{T}}\dot{\theta}/2$, the most preferred RMP performance index (2.2), and extension-II performance index (2.12)] on PUMA560 robot manipulator with its end-effector tracking a straight-line path, respectively

C.2.2 Straight-Line Path Tracking

In this example, the end-effector of the PUMA560 manipulator is expected to move forward and then backward along a straight-line segment. First, let us consider the situation of no drift-free criterion. As we see from the upper graph of Fig. C.4, the

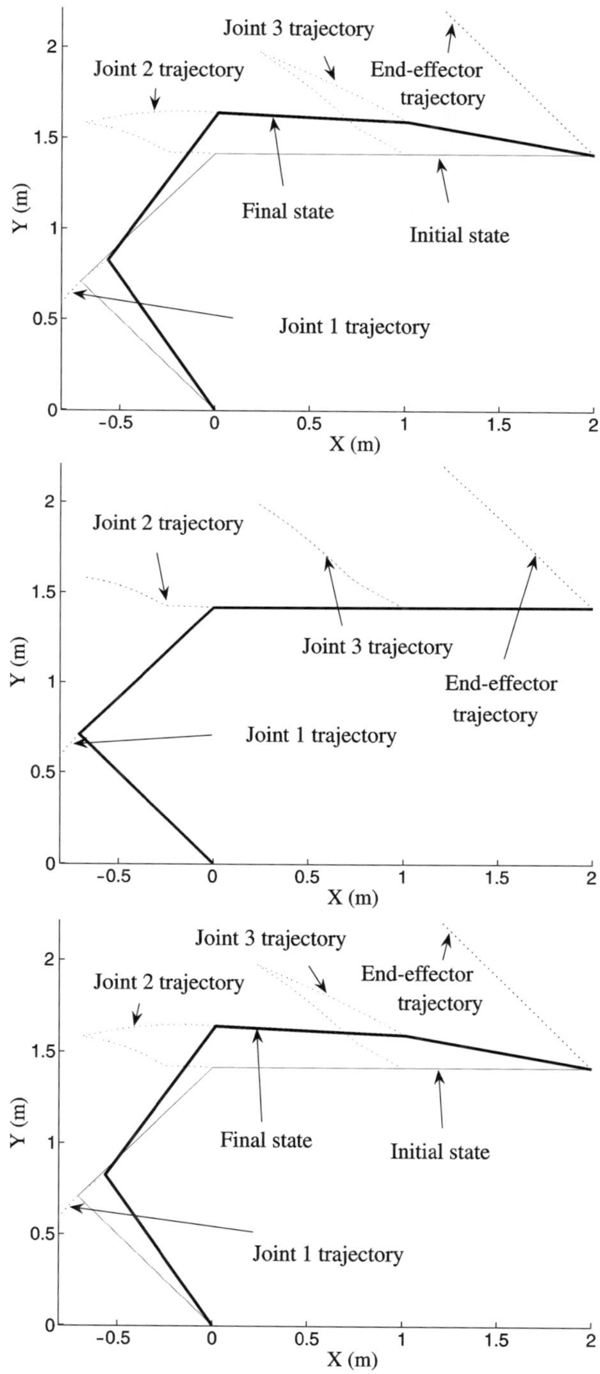

Fig. C.5 Performances of different indices [i.e., minimum-velocity performance index $\dot{\theta}^{\mathrm{T}}\dot{\theta}/2$, the most preferred RMP performance index (2.2), and extension-II performance index (2.12)] on four-link planar manipulator with its end-effector tracking a straight-line path, respectively

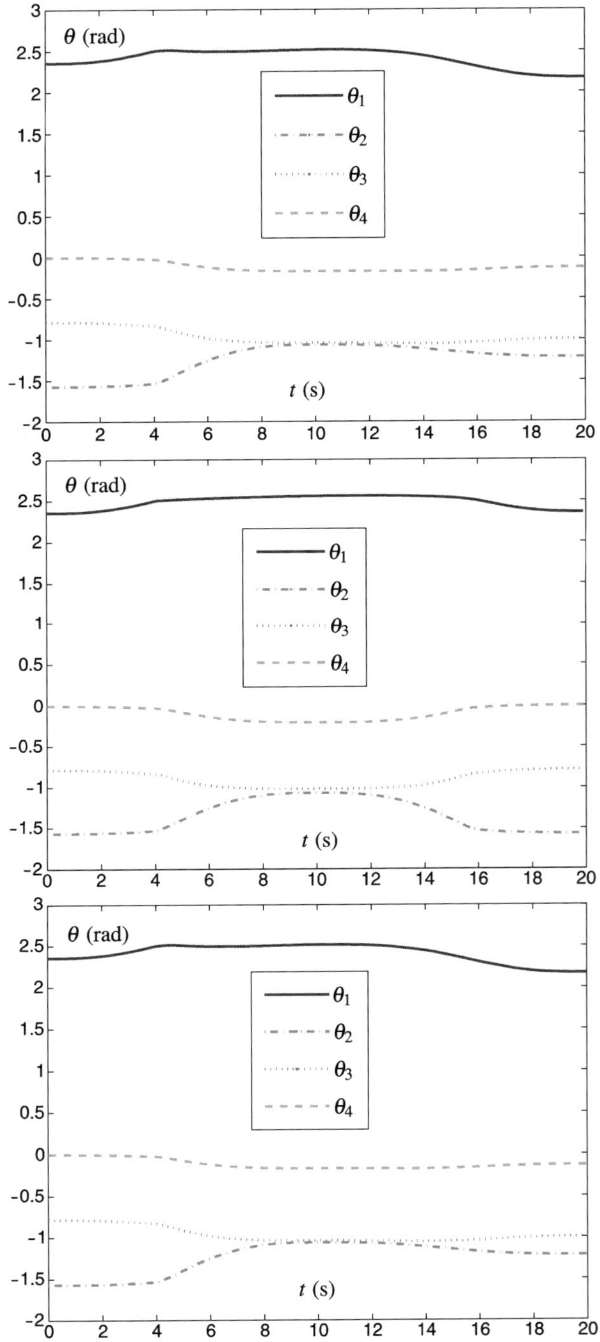

Fig. C.6 Joint profiles of four-link planar manipulator with its end-effector tracking a straight-line path, synthesized by different indices [i.e., minimum-velocity performance index $\dot{\theta}^{\mathrm{T}}\dot{\theta}/2$, the most preferred RMP performance index (2.2), and extension-II performance index (2.12)], respectively solution (actually, as we have observed, many of the inverse-kinematics solutions) has the joint-angle drift phenomenon.

Second, let us consider the inclusion of the most preferred RMP performance index (2.2). As we see from the middle graph of Fig. C.4, the solution is now repetitive (or to say, cyclic), in view of the fact that the final PUMA560 state coincides with its initial state.

Third, for comparison and illustration, we take the extension-II performance index (2.12) into account, and the corresponding simulation result is shown in the lower graph of Fig. C.4. Evidently, we see again that the solution synthesized by (2.12) is not repetitive, and, in addition, it is similar to that synthesized by $\dot{\theta}^{\mathrm{T}}\dot{\theta}/2$.

C.3 Four-Link Planar Robot Manipulator

Different from the above two examples of using three-dimensional robot manipulators [wherein PA10 is kinematically redundant, while PUMA560 could be functionally redundant (i.e., redundant for specific end-effector tasks)], in this section we take a planar robot manipulator (i.e., the four-link planar robot manipulator) as another example [3, 5].

Specifically, the upper graph of Fig. C.5 illustrates the nonrepetitive simulation result synthesized by minimum-velocity performance index $\dot{\theta}^{\mathrm{T}}\dot{\theta}/2$ (and with joint physical limits considered as well), the middle graph of Fig. C.5 illustrates the repetitive simulation result synthesized by the most preferred RMP performance index (2.2), and the lower graph of Fig. C.5 illustrates the nonrepetitive simulation result synthesized by extension-II performance index (2.12). In addition, their corresponding joint profiles are shown in Fig. C.6.

These simulation results demonstrate again that the most preferred RMP scheme (2.2)–(2.5) is effective on the repetitive robot-motion planning, whereas its simple extensions [e.g., the extension-II optimization scheme (2.12)–(2.15)] may not work applicably. Besides, it is worth pointing out that our further consideration and simulation results, as shown in Sect. 2.3.3 and in the next appendix, show that using linear performance index $c^{\mathrm{T}}\dot{\theta}$ (or to say, extension-III performance index $c^{\mathrm{T}}\dot{\theta}$) could achieve the RMP objective of initial and final states being equal, but "slowly" and with a chattering phenomenon. For details, please see Sect. 2.3.3 in the next appendix.

References

1. Chen K, Guo D, Tan Z, Yang Z, Zhang Y (2008) Cyclic motion planning of redundant robot arms: simple extension of performance index may not work. In: Proceedings of the 2nd international symposium on intelligent information technology application, pp 635–639
2. Zhang Y, Tan Z, Yang Z, Lv X, Chen K (2008) A simplified LVI-based primal–dual neural network for repetitive motion planning of PA10 robot manipulator starting from different initial states. In: Proceedings of international joint conference on neural networks, pp 19–24

3. Zhang Y (2002) Analysis and design of recurrent neural networks and their applications to control and robotic systems. PhD dissertation, Chinese University of Hong Kong, Hong Kong
4. Zhang Y, Ge SS, Lee TH (2004) A unified quadratic programming based dynamical system approach to joint torque optimization of physically constrained redundant manipulators. IEEE Trans Syst Man Cybern, Part B, Cybern 34(5):2126–2132
5. Cheng F, Chen T, Sun Y (1994) Resolving manipulator redundancy under inequality constraints. IEEE J Robot Autom 10(1):65–71
6. Zhang Y, Wang J (2002) A dual neural network for constrained torque optimization of kinematically redundant manipulators. IEEE Trans Syst Man Cybern, Part B, Cybern 32(5):654–662

Appendix D
Simulations of Extension-III Scheme

In this appendix, computer simulations are performed based on three different types of robot arms to verify the efficacy of linear performance index (2.17) and its corresponding LP-based scheme (2.17)–(2.20) on robots' repetitive redundancy resolution [1].

D.1 PUMA560 Robot Manipulator

In this section, computer simulations are conducted based on the 6-DOF PUMA560 robot manipulator again and with positioning considered only [2–4]. Two simulation examples are shown as follows, i.e., a circular-path tracking example and a straight-line path tracking example.

D.1.1 Circular Path Tracking

In this circular example, the motion trajectory of the PUMA560 end-effector is expected to be a circular path with radius 0.1 m and the revolute angle about X-axis being $\pi/6$ rad. The motion-task duration is 10 s, and the initial joint state $\theta(0) = [0, 0, 0, 0, 0, 0]^{\mathrm{T}}$ rad.

As seen from the upper graph of Fig. D.1, the end-effector of the PUMA560 robot manipulator moves along a circle in the three-dimensional workspace, which is sufficiently close to the desired circular path (with the maximal position error less than 3.469×10^{-5} m). In addition, as seen from the middle graph of Fig. D.1, the final state of the solution $\theta(t)$ is equal to its initial state; mathematically, $\theta_1(0) = \theta_1(10)$, $\theta_2(0) = \theta_2(10)$, $\theta_3(0) = \theta_3(10)$, $\theta_4(0) = \theta_4(10)$, $\theta_5(0) = \theta_5(10)$, and $\theta_6(0) = \theta_6(10)$. Hence, the LP-based scheme (2.17)–(2.20) could be considered as a successful repetitive-motion-planning scheme. In other words, the linear performance index (2.17) appears to be effective for the PUMA560 robot manipulator

Y. Zhang, Z. Zhang, *Repetitive Motion Planning and Control of Redundant Robot Manipulators*, DOI 10.1007/978-3-642-37518-7,

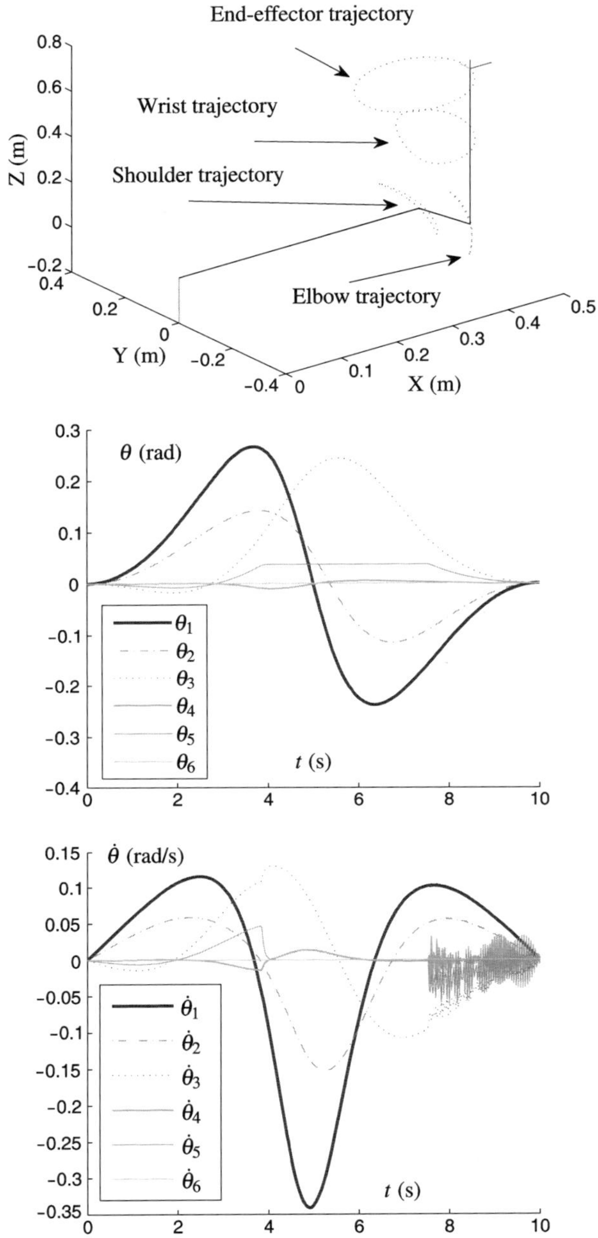

Fig. D.1 PUMA560 end-effector tracking a circular path, which is synthesized by linear performance index (2.17) and its corresponding LP-based scheme (2.17)–(2.20)

working with a repetitive motion requirement. However, by looking into the detail and intermediate results, we may see from the lower graph of Fig. D.1 that the solution velocity $\dot{\theta}(t)$ has chattering phenomena, which may wear down the PUMA560 manipulator considerably. So, this LP-based scheme, though successful, might be less efficient, less desirable, and less applicable in engineering applications.

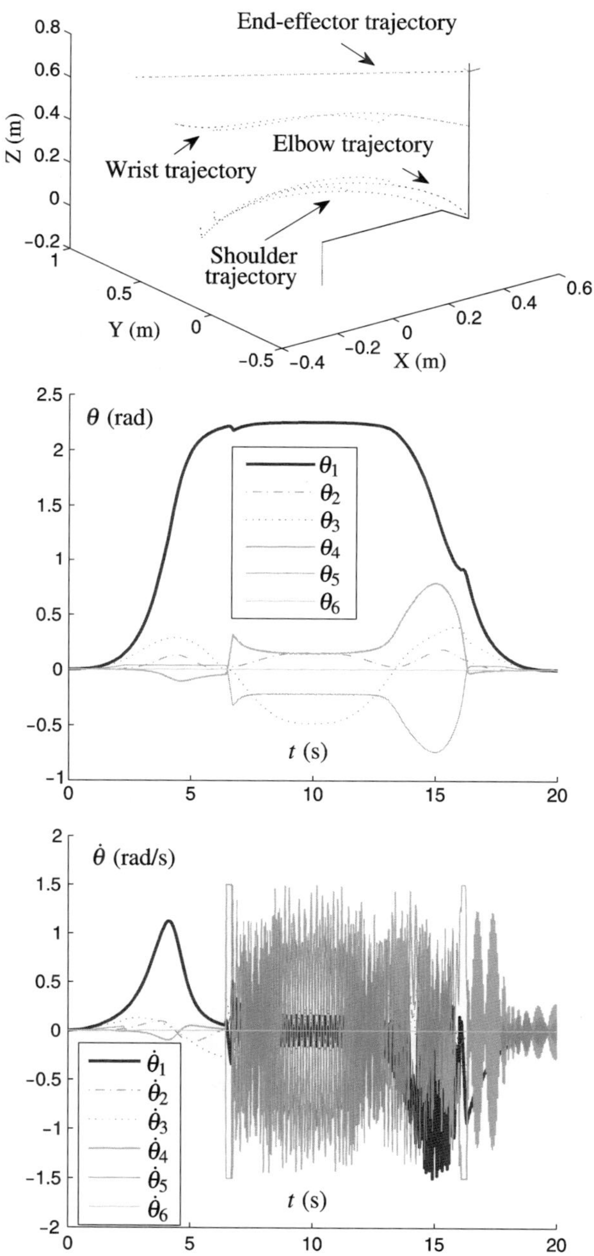

Fig. D.2 PUMA560 end-effector tracking a straight-line path, which is synthesized by linear performance index (2.17) and its corresponding LP-based scheme (2.17)–(2.20)

D.1.2 Straight-Line Path Tracking

In this second example of the PUMA560 robot manipulator, the motion trajectory of the end-effector is expected to move forward and then backward along a straight-line segment. The duration of this straight-line path tracking task is

20 s. In addition, such a line-segment is of length 2.25 m, and the initial state $\theta(0) = [0, 0, 0, 0, 0, 0]^{\mathrm{T}}$ rad. Angles of the desired straight-line segment made with XY, YZ, and ZX planes are 0 rad, $\pi/4$ rad, and $\pi/4$ rad, respectively.

The upper graph of Fig. D.2 illustrates the LP-synthesized simulation-trajectory of the PUMA560 end-effector, which is very close to the desired straight-line segment in the three-dimensional workspace (with the maximal position error less than 3.780×10^{-4} m). Correspondingly, the joint vector θ and joint-velocity vector $\dot{\theta}$ synthesized by the LP-based scheme (2.17)–(2.20) are shown in the middle and lower graphs of Fig. D.2, respectively. According to the upper and middle graphs of Fig. D.2, the LP-based scheme could achieve the repetitive-motion-planning purpose in the sense that the final and initial states of the PUMA560 robot manipulator coincide very well with each other. However, as we see form the lower graph of Fig. D.2, there exist chattering phenomena again in the joint velocity! Therefore, the solution (including θ and $\dot{\theta}$) of the LP-based scheme appears to be inapplicable for practical repetitive-motion (or to say, cyclic-motion) tasks because of its oscillation.

In summary, the circular path example and straight-line path example have both demonstrated that the linear performance index (2.17) and its corresponding LP-based scheme (2.17)–(2.20) could achieve the repetitive-motion-planning purpose, but there might exist undesired chattering phenomena in the joint velocity [i.e., $\dot{\theta}(t)$] of the corresponding solution $\theta(t)$. Thus, this LP-based scheme may not be applicable for engineering applications. A question following this summary is "How often or how generally does the chattering phenomenon happen when we use the linear performance index and/or the linear-programming technique?"

D.2 PA10 Robot Manipulator

To address the above question, further computer simulations are conducted and presented in this section for the situation of the PA10 robot manipulator [2, 5, 6].

D.2.1 Circular Path Tracking

In the simulation of this subsection, the motion trajectory of the PA10 end-effector is expected to be a circle with radius 0.2 m and the revolute angle about X-axis being $\pi/6$ rad. The task duration is 10 s, and the initial joint state $\theta(0) = [0, \pi/4, 0, 2\pi/3, 0, -\pi/4, 0]^{\mathrm{T}}$ rad.

The simulated results are shown in Fig. D.3. As seen from the upper graph of Fig. D.3, the final and initial PA10 states coincide well with each other. Meanwhile, from the middle graph of Fig. D.3 we see that the final values of joint variables [i.e., the elements of $\theta(t)$] are equal to their initial values. The solution is thus repetitive in the usual joint-angle sense. Unfortunately, the joint-velocity profiles exist oscillation, as shown in the lower graph of Fig. D.3, which is evidently an unwilling

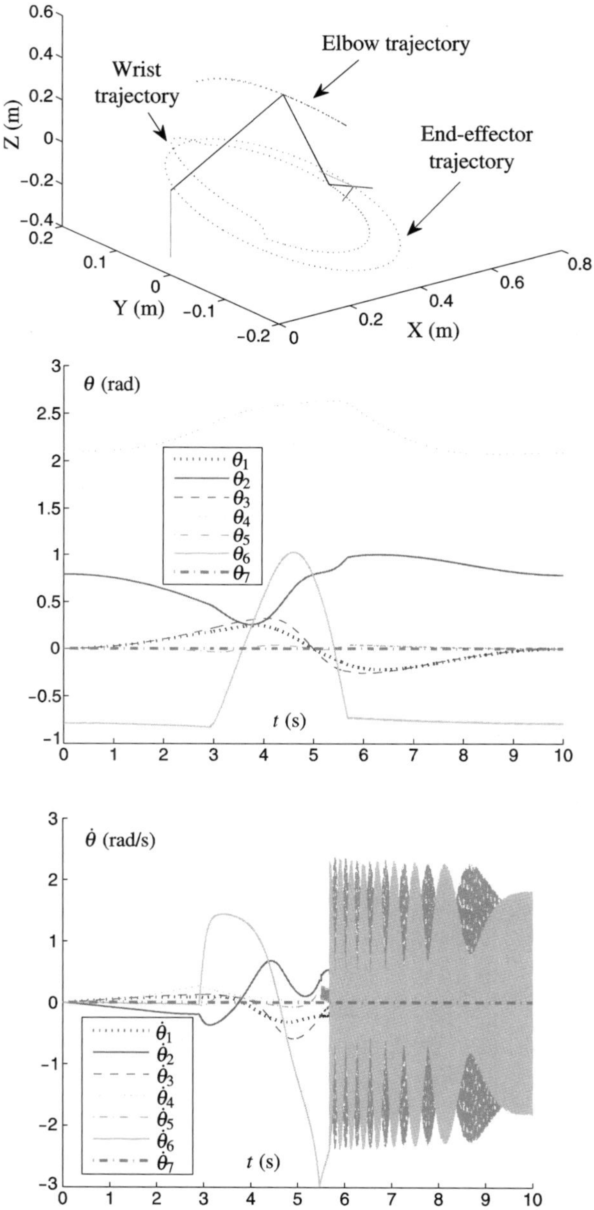

Fig. D.3 PA10 end-effector tracking a circular path, which is synthesized by linear performance index (2.17) and its corresponding LP-based scheme (2.17)–(2.20)

phenomenon in engineering applications. Thus, the linear performance index (2.17) and its corresponding LP-based scheme (2.17)–(2.20) are not applicable again for the PA10 robot task due to such chattering phenomena.

D.2.2 Straight-Line Path Tracking

In this second simulation of the PA10 robot manipulator, the motion trajectory of the end-effector is to move forward and then backward along a straight-line segment. The task duration is 7 s, the line-segment length is 2.5 m, and the initial state $\theta(0) = [\pi/4, \pi/4, \pi/4, \pi/4, \pi/4, \pi/4, 0]^{\mathrm{T}}$ rad. The angles of the desired straight-line segment making with XY, YZ, and ZX planes are $\pi/4$ rad, $\pi/6$ rad, and $\pi/6$ rad, respectively.

The upper graph of Fig. D.4 shows the PA10 motion trajectories. Simply put, the solution is repetitive (or to say, cyclic, conservative) in the sense that its final state equals the initial state. In addition, the profiles of joint-angle variables in the middle graph of Fig. D.4 substantiate as well that the LP-based scheme (2.17)–(2.20) could achieve the RMP purpose. However, as seen again from the lower graph of Fig. D.4, the linear performance index (2.17) and its LP-based scheme (2.17)–(2.20) may not be suitable for engineering applications because of the chattering phenomena.

In summary, the circular path simulation and the straight-line path simulation based on the PA10 robot manipulator have both demonstrated that the observation and conclusion about the LP-based scheme are quite similar to those based on the PUMA560 robot manipulator. That is, the LP-based scheme (2.17)–(2.20) could achieve the repetitive-motion-planning purpose, but its solution [mainly $\dot{\theta}(t)$] generally has chattering phenomena.

Furthermore, for performance comparison with the most preferred quadratic RMP performance index (2.2) and its QP-based scheme (2.2)–(2.5), see Chaps. 9 and 10, especially Figs. 9.5 and 9.6 for PUMA560 and Figs. 10.3, 10.4, 10.8, and 10.9 for PA10. In addition to those, Figs. D.5 and D.6 are presented for readers' convenience, which show the corresponding results of $\dot{\theta}(t)$ synthesized by the most preferred quadratic RMP performance index (2.2) and its QP-based scheme (2.2)–(2.5). From these two QP-synthesized figures (i.e., Figs. D.5 and D.6) we see that the chattering phenomenon vanishes. Besides, it is worth mentioning that the robot simulation related to the LP-based scheme appears to be much slower [i.e., taking longer time (e.g., a few hours) to complete it] than that related to the QP-based scheme. So, we may conclude that the QP-based scheme is much more efficient and applicable for robot applications than the LP-based scheme.

D.3 Four-Link Planar Robot Manipulator

Different from the above four examples of using two three-dimensional robot manipulators (i.e., kinematically-redundant PA10 robot manipulator and functionally-redundant PUMA560 robot manipulator), in this example we take the four-link planar robot manipulator [2, 7] as another example.

Figures D.7 and D.8 illustrate the computer-simulation results of the four-link planar manipulator's end-effector tracking a straight-line path, which are synthesized by using the LP-based scheme (2.17)–(2.20) and the QP-based scheme (2.2)–(2.5), respectively. Specifically speaking, the upper graphs of Fig. D.7 and Fig. D.8

Fig. D.4 PA10 end-effector tracking a straight-line path, which is synthesized by linear performance index (2.17) and its corresponding LP-based scheme (2.17)–(2.20)

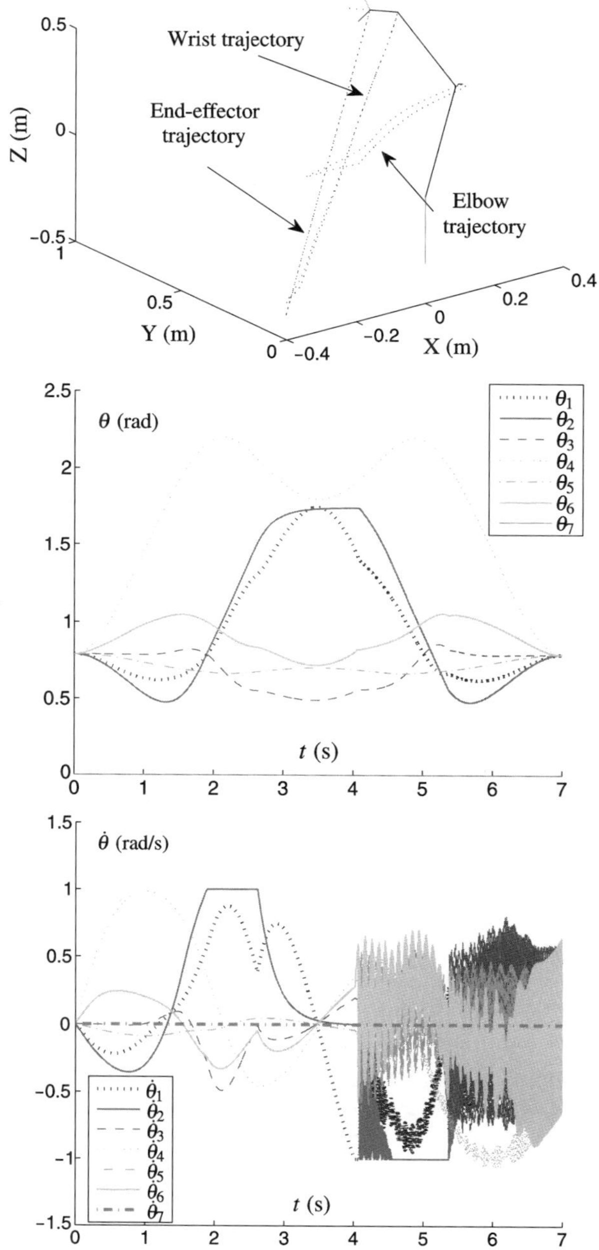

show the motion trajectories of the four-link planar robot manipulator, the middle graphs of Fig. D.7 and Fig. D.8 show the trajectories of joint-angle variables of the four-link planar robot manipulator, and the lower graphs of Fig. D.7 and Fig. D.8 show the simulated joint velocities. These simulation results demonstrate again that

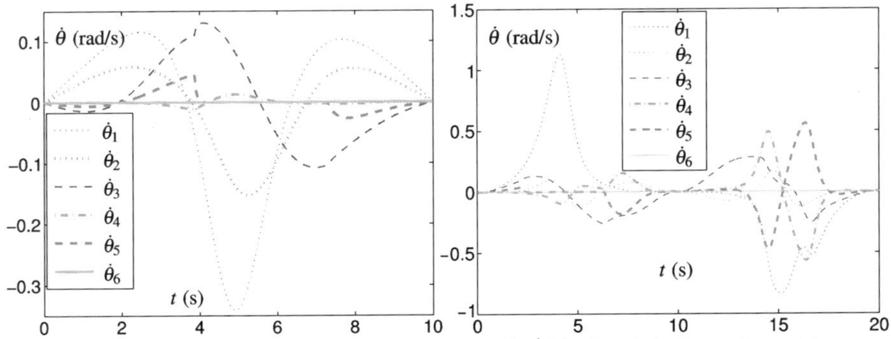

(a) $\dot{\theta}(t)$ of circular-path tracking (b) $\dot{\theta}(t)$ of straight-line path tracking

Fig. D.5 PUMA560 joint-velocity profiles synthesized by the most preferred quadratic RMP performance index (2.2) and its QP-based scheme (2.2)–(2.5), in addition to Figs. 9.5 and 9.6

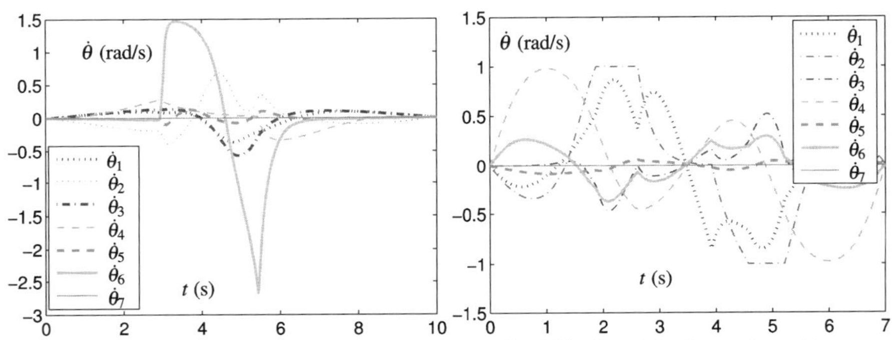

(a) $\dot{\theta}(t)$ of circular-path tracking (b) $\dot{\theta}(t)$ of straight-line path tracking

Fig. D.6 PA10 joint-velocity profiles synthesized by the quadratic RMP performance index (2.2) and its QP-based scheme (2.2)–(2.5), in addition to Figs. 10.3, 10.4, 10.8, and 10.9

the LP-based scheme (2.17)–(2.20) could achieve the repetitive-motion-planning purpose for such a planar redundant robot manipulator, but its performance is less acceptable for real applications due to the chattering phenomena. On the other hand, the QP-based scheme (2.2)–(2.5) can achieve the repetitive-motion-planning purpose for various types of redundant robot arms (including the four-link planar robot manipulator), and its performance appears to be more smooth, more efficient, and more applicable for robot applications (as compared to the LP-based scheme) [2, 5–7].

In summary, all of the computer-simulation results based on different types of redundant robot manipulators have demonstrated that the LP-based scheme (2.17)–(2.20) may quite often introduce the chattering phenomena onto the resolved joint velocity, though such an LP-based scheme could achieve the repetitive-motion-planning purpose. This summary is compared with the more-successful QP-based scheme (2.2)–(2.5). Furthermore, the simulation results of Appendix C have shown

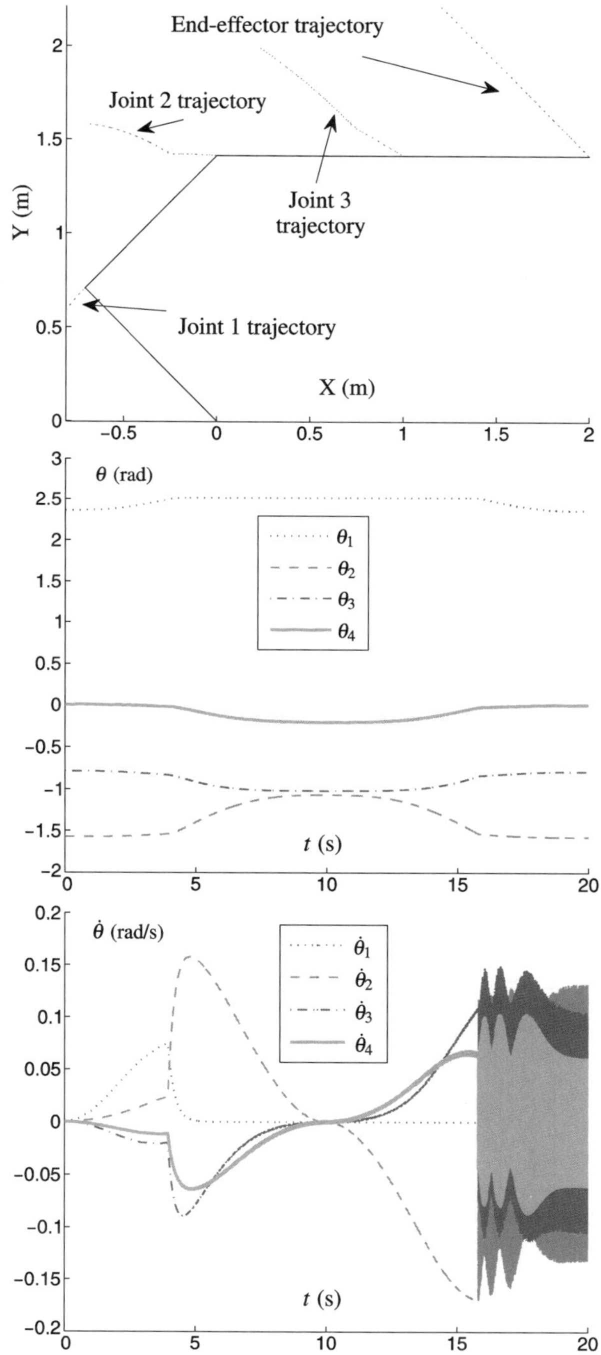

Fig. D.7 Four-link planar manipulator's end-effector tracking a straight-line path as synthesized by linear performance index (2.17) and its LP-based scheme (2.17)–(2.20)

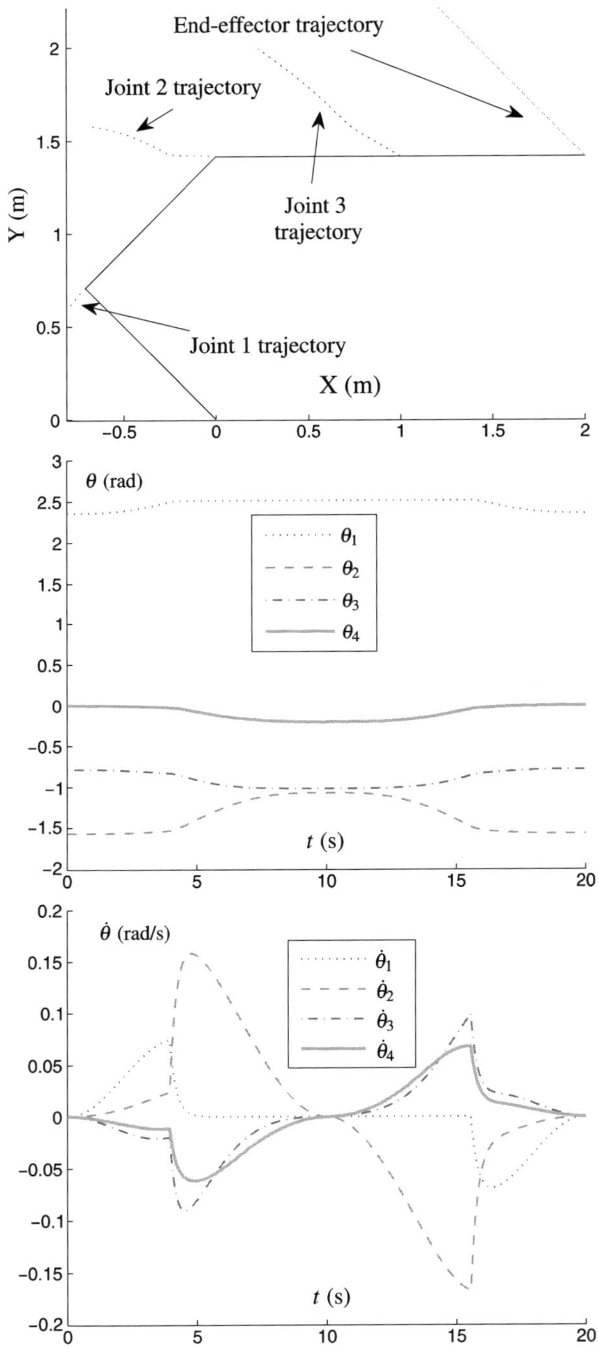

Fig. D.8 Four-link planar manipulator's end-effector tracking a straight-line path as synthesized by the most preferred quadratic RMP performance index (2.2) and its QP-based scheme (2.2)–(2.5)

that using the pure quadratic performance index $\dot{\theta}^{\mathrm{T}} W \dot{\theta}/2$ in place of (2.2) with $W := I$ or $W := \|\theta - \theta(0)\|_2^2 I$ could not achieve the RMP purpose for redundant robots. This indicates that the linear part of (2.2) [or to say, the linear performance index $\lambda(\theta - \theta(0))^{\mathrm{T}}\dot{\theta}$ in (2.17)], though introducing the chattering phenomena when used alone, actually plays an important role in the success of the most preferred quadratic-form RMP scheme (2.2)–(2.5) on robot manipulators.

References

1. Zhang Y, Guo D, Tan Z, Yang Z (2009) Linear programming versus quadratic programming in robots' repetitive redundancy resolution: a chattering phenomenon investigation. In: Proceedings of the 4th IEEE conference on industrial electronics and applications, pp 2822–2827
2. Zhang Y (2002) Analysis and design of recurrent neural networks and their applications to control and robotic systems. PhD dissertation, Chinese University of Hong Kong, Hong Kong
3. Zhang Y, Ge SS, Lee TH (2004) A unified quadratic programming based dynamical system approach to joint torque optimization of physically constrained redundant manipulators. IEEE Trans Syst Man Cybern, Part B, Cybern 34(5):2126–2132
4. Zhang Y, Wang J (2002) A dual neural network for constrained torque optimization of kinematically redundant manipulators. IEEE Trans Syst Man Cybern, Part B, Cybern 32(5):654–662
5. Zhang Y, Tan Z, Yang Z, Lv X, Chen K (2008) A simplified LVI-based primal–dual neural network for repetitive motion planning of PA10 robot manipulator starting from different initial states. In: Proceedings of international joint conference on neural networks, pp 19–24
6. Zhang Y, Wang J, Xia Y (2003) A dual neural network for redundancy resolution of kinematically redundant manipulators subject to joint limits and joint velocity limits. IEEE Trans Neural Netw 14(3):658–667
7. Cheng F, Chen T, Sun Y (1994) Resolving manipulator redundancy under inequality constraints. IEEE J Robot Autom 10(1):65–71

Glossary

kinematically redundant manipulator A kinematically redundant manipulator is a robot manipulator having more kinematic degrees of freedom (DOF) than required for its end-effector's position and orientation.

pseudoinverse-type solution It includes a minimum-norm particular solution and a homogeneous solution.

nonrepetitive motion problem When the end-effector traces a closed path in its workspace, the joint angles may not return to their initial values after completing the end-effector task.

primal neural network A recurrent neural network is called primal neural network if the network dynamic equation and implementation only use the original (also termed primal) decision variables (here, the original/primal decision vector is $\hat{x}$).

nonprimal neural network A recurrent neural network is called nonprimal neural network if the network dynamic equation and implementation use a group of auxiliary decision variables, in addition to using the original/primal variables or not.

primal–dual neural network A recurrent neural network is called primal–dual neural network if the network dynamic equation and implementation use a group of dual decision variables, in addition to using the original/primal ones.

dual neural network A recurrent neural network is called dual neural network if the network dynamic equation and implementation only use a group of dual decision variables.

global convergence of a recurrent neural network A recurrent neural network is said to be globally convergent if starting from any initial point $\hat{x}(t_0)$ taken in the whole associated Euclidean space, every state trajectory of the recurrent neural network converges to an equilibrium point $\hat{x}^*$ that depends on the initial state of the trajectory. Note that the initial time instant $t_0 \geqslant 0$ and there might exist many nonisolated equilibrium points $\hat{x}^*$.

global exponential convergence of a recurrent neural network A recurrent neural network is said to be globally exponentially convergent if every trajectory starting from any initial point $\hat{x}(t_0)$ satisfies that, for all $t \geqslant t_0 \geqslant 0$, $\|\hat{x}(t) - \hat{x}^*\| \leqslant \eta\|\hat{x}(t_0) - \hat{x}^*\| \exp(-\lambda(t - t_0))$, where η and λ are positive constants, $\hat{x}^*$ is an

Y. Zhang, Z. Zhang, *Repetitive Motion Planning and Control of Redundant Robot Manipulators*, DOI 10.1007/978-3-642-37518-7,

equilibrium point depending on initial state $\hat{x}(t_0)$, and the symbol $\|\cdot\|$ denotes the two-norm of a vector unless specified otherwise.

stability of an equilibrium point An equilibrium point $\hat{x}^*$ of a recurrent neural network is locally stable if every trajectory starting from the initial condition $\hat{x}(t_0)$ near the equilibrium point $\hat{x}^*$ stays near that equilibrium point $\hat{x}^*$. In addition, an equilibrium point $\hat{x}^*$ of a recurrent neural network is globally stable if every trajectory starting from any initial condition $\hat{x}(t_0)$ stays near that equilibrium point $\hat{x}^*$.

Printed by Printforce, the Netherlands